CHEMICAL ANALYSIS
OF THE ENVIRONMENT
AND OTHER MODERN TECHNIQUES

PROGRESS IN
ANALYTICAL CHEMISTRY
Based upon the Eastern Analytical Symposia

Series Editors:
Ivor L. Simmons
M&T Chemicals, Inc., Rahway, N.J.

and Galen W. Ewing
Seton Hall University, South Orange, New Jersey

Volume 1
H. van Olphen and W. Parrish
X-RAY AND ELECTRON METHODS OF ANALYSIS
Selected papers from the 1966 Eastern Analytical Symposium

Volume 2
E. M. Murt and W. G. Guldner
PHYSICAL MEASUREMENT AND ANALYSIS OF THIN FILMS
Selected papers from the 1967 Eastern Analytical Symposium

Volume 3
K. M. Earle and A. J. Tousimis
X-RAY AND ELECTRON PROBE ANALYSIS
IN BIOMEDICAL RESEARCH
Selected papers from the 1967 Eastern Analytical Symposium

Volume 4
C. H. Orr and J. A. Norris
COMPUTERS IN ANALYTICAL CHEMISTRY
Selected papers from the 1968 Eastern Analytical Symposium

Volume 5
S. Ahuja, E. M. Cohen, T. J. Kneip, J. L. Lambert, and G. Zweig
CHEMICAL ANALYSIS OF THE ENVIRONMENT AND
OTHER MODERN TECHNIQUES
Selected papers from the 1971 Eastern Analytical Symposium

PROGRESS IN ANALYTICAL CHEMISTRY
VOLUME 5

CHEMICAL ANALYSIS OF THE ENVIRONMENT
AND OTHER MODERN TECHNIQUES

Edited by

Sut Ahuja
CIBA—Geigy Corporation
Summit, New Jersey

Edward M. Cohen
Merck, Sharp and Dohme Research Laboratories
West Point, Pennsylvania

Theo. J. Kneip
Laboratory for Environmental Studies
New York University Medical Center
New York, New York

Jack L. Lambert
Department of Chemistry
Kansas State University
Manhattan, Kansas

Gunter Zweig
Syracuse University Research Corporation
University Heights
Syracuse, New York

PLENUM PRESS • NEW YORK–LONDON • 1973

Library of Congress Catalog Card Number 73-82575

ISBN 978-1-4684-7247-9   ISBN 978-1-4684-7245-5 (eBook)
DOI 10.1007/978-1-4684-7245-5

© *1973 Plenum Press, New York*
Softcover reprint of the hardcover 1st edition 1973

A Division of Plenum Publishing Corporation
227 West 17th Street, New York, New York 10011

United Kingdom edition published by Plenum Press, London
A Division of Plenum Publishing Company, Ltd.
Davis House (4th Floor), 8 Scrubs Lane, Harlesden, London NW10 6SE, England

PREFACE

With the rise in general awareness of the
effects of trace chemicals in the environment on
man's health, it has been realized that traditional
methods of analysis are often inadequate. Reliable
analyses are needed in the fractional parts-per-
million range of contaminants in condensed phases,
and of the order of micrograms per cubic meter in
air. Trying to get meaningful answers regarding
such minute amounts raises cogent problems in all
stages of an analysis.

It is most appropriate, therefore, that the
1971 Eastern Analytical Symposium should have four
half-day sessions devoted to this general field.
Two of these, entitled "Trace Metals in the Environment," were assembled by Dr. Kneip, one on
"Pesticides in the Environment: Recently Discovered
Analytical Problems," by Dr. Zweig, and one on "The
Determination of Anions in Water," by Dr. Lambert.
Together, these reports furnish a fairly complete
picture of the present state of environmental analysis.

The remainder of this volume is devoted to
pharmaceutical analysis, a diversified field in
which nearly all analytical methods find a place.
Partly because of this multiplicity of techniques,
and partly due to the large number of samples
which must be examined in connection with the manufacture, biological testing, and clinical application of pharmaceutical preparations, this area is
particularly appropriate for the introduction of
automation. The objective, broadly, is to speed up
multiple analyses without the sacrifice of accuracy.

The 1971 Eastern Analytical Symposium was fortunate to be able to present two sessions in this field: "Current Topics in Pharmaceutical Analysis," under the chairmanship of Dr. Cohen, and "Automated Analysis," chaired by Dr. Ahuja. The Editors wish to take this occasion to thank the session chairmen and the individual authors for their contributions to this Symposium.

Galen W. Ewing

Ivor L. Simmons

CONTENTS

TRACE METALS IN THE ENVIRONMENT

Edited by Theo. J. Kneip

Pitfalls in the Determination of Environmental
Trace Metals 3
David Hume

Atmospheric Trace Metal Studies 17
Herbert L. Volchok and Donald C. Bogen

Trace Metal Concentration Factors in Aquatic
Ecosystems 43
Theo. J. Kneip and Gerald J. Lauer

Human Trace Metal Burdens: Pathways, Uptake,
and Loss 63
Gwyneth Howells

Relation of Trace Metals to Human Health
Effects 81
Henry A. Schroeder and Dan K. Darrow

PESTICIDES IN THE ENVIRONMENT: RECENTLY
DISCOVERED ANALYTICAL PROBLEMS

Edited by Gunter Zweig

Introduction 107
Gunter Zweig

Polychlorinated Biphenyls: Their Potential
Interference with Pesticide Residue
Analyses and Analytical Status 109
Jeffrey L. Lincer

Analysis of Pesticides in Air 133
    Bill Compton

The Significance of Pesticide Contaminants 153
    J. R. Plimmer

Gas Chromatographic Analysis of Pesticide
        Residues Containing Phosphorus and/or
        Sulfur with Flame Photometric Detection
        and Some Ancillary Techniques for
        Verifying their Identities 175
    M. C. Bowman

THE DETERMINATION OF ANIONS IN WATER

Edited by Jack L. Lambert

Some Partially-Solved Problems in Anion Analysis . . . 195
    Jack L. Lambert

Spectrophotometric, Spectrofluorometric, and
        Atomic Absorption Spectrometric
        Methods for the Determination of
        Anions 201
    David F. Boltz

Fluoride Analysis in Seawater and in Other
        Complex Natural Waters Using an
        Ion-Selective Electrode --
        Techniques, Potentialities,
        Limitations 229
    Theodore B. Warner

Analytical Techniques for Nutrient Anions
        and Other Pertinent Substances
        in the Water Environment 241
    Leonard L. Ciaccio

CURRENT TOPICS IN PHARMACEUTICAL ANALYSIS

Edited by Edward M. Cohen

Analytical Requirements of Automated
        Pharmaceutical Analysis 275
    Andres Ferrari

A Look Ahead Towards Possible Future
        Analytical Requirements by FDA 277
    Daniel Banes

Data and Information Handling Systems in the
        Pharmaceutical, Analytical, and
        Quality Control Environments 285
    H. Stelmach

Panel Discussion

    Semantics in Specifications for Drugs 301
      Lester Chafetz

    What Problems Does One Introduce When
            Different Criteria for Accept-
            ability of Unfinished Actives
            vs. the Formulated Actives Are
            Utilized? 305
        G. J. Papariello

    Analytical Methods for Stability Samples--
            What Should They Tell you? 309
        Bernard Z. Senkowski

    Tests to Monitor the Acceptability of
            Raw Materials -- Are They
            Adequate? 313
        Anthony J. Taraszka

AUTOMATED ANALYSIS

Edited by Sut Ahuja

Automation in Microchemistry: Automatic CHN
        Analyzer Case History 319
    Grant M. Gustin

Methodology Problems in Automated Analyses 337
    S. Ahuja

Growth Path for Computers in Automated Analyses . . . 357
    Henderson Cole

I N D E X 377

Trace Metals in the Environment

Edited by **Theo. J. Kneip**

PITFALLS IN THE DETERMINATION OF ENVIRONMENTAL TRACE METALS

David N. Hume

Department of Chemistry
Massachusetts Institute of Technology
Cambridge, Mass. 02139

INTRODUCTION

The growing realization of the importance of even extremely small amounts of metals in the environment has led to an ever increasing demand for determination of these metals at trace-level concentrations. The resulting flood of data has evoked a mixed response from members of the public, scientific and otherwise. Some accept the results unhesitatingly, particularly if in conformity with their prejudices, and make them the basis for legislative action, prosecution, indifference or panic. Others deny that quantities so small can be measured with any degree of meaning, while still others are selectively skeptical, or simply bewildered. In recognition of this state of affairs and of the importance of reliable and interpretable data in the many fundamental scientific and social problems related to the presence of trace metals in the environment, I would like to examine with you the present state of the art and its implications.

RELIABILITY OF TRACE METHODS

In trace level analysis we are dealing with concentrations no greater than the parts per million level and more often today with parts per billion or less. Our concern is with all phases of the environment: air, water, solid

earth and the biosphere, and includes indigenous metals as
well as those deriving from pollution. In order to measure
at the parts per billion (i.e., micrograms per kilogram)
level, methods of high sensitivity and selectivity are needed.
They must be applicable to real-world samples, accurate and
reliable, and preferably they should also be convenient and
economical. According to the literature there are several
methods which meet all the essential requirements and indeed
we find widespread adoption of atomic emission and absorption spectroscopy and colorimetry as well as significant use
of neutron activation, electrochemical methods and mass
spectrometry. The authors of many papers assure us that the
necessary precision and accuracy (usually quoted as in the
range of 5% at worst down to a few tenths of a percent under
the most favorable conditions) is readily achieved in their
laboratories, and we gain the impression that while minor
adjustments may have to be worked out for applications to
specific systems, the major problems have been identified
and solved.

Now it is always prudent to be somewhat skeptical of
broad claims made by enthusiastic authors. Such claims are,
in general, made honestly, but not on the basis of particularly wide experience. In order to learn what a method might
do for you, it is best to look at what it does do in the hands
of a cross section of competent workers under circumstances
where they have no idea of what the correct value might be.
This is best accomplished by distributing portions of a
carefully prepared and preserved unknown sample to a number
of qualified analysts to run independently and report under
code for comparison. A number of such studies have been
made, particularly for the determination of trace metals in
water, and it is rather instructive to look at some of the
results.

In Table 1 is summarized some of the findings in a
cooperative interlaboratory study of the determination of
trace metals in water using conventional, widely accepted
colorimetric methods (1). The 79 cooperating groups, which
were all participating voluntarily, were professional water
laboratories representing federal, state, municipal, academic,
industrial and private organizations. Each had the same
synthetic sample, which was run in triplicate, and the
averages were used for comparisons between laboratories.
Results which were clearly affected by some gross error,
such as a factor of ten in dilution, were omitted from the

TABLE 1

STANDARD COLORIMETRIC TRACE METAL PROCEDURES FOR WATER
RESULTS OF INTERLABORATORY COMPARATIVE STUDY

Element	Labs	Added p.p.m.	Ave Found	Range p.p.m.	Std Dev %	Error %
Ag	14*	0.15	0.049	0.00-0.10	20	-67
Al	44	0.50	0.400	0.00-0.81	30	-20
Cd	44	0.05	0.053	0.02-0.11	20	+ 6
Cr	31	0.11	0.092	0.00-0.21	36	-16
Cu	25	0.47	0.514	0.26-0.72	19	+ 9
Fe	40	0.30	0.337	0.08-0.53	30	+12
Mn	33*	0.12	0.118	0.03-0.21	42	- 2
Pb	42*	0.07	0.076	0.00-0.20	43	+ 9
Zn	46	0.65	0.818	0.34-1.3	23	+26

comparisons, and an asterisk indicates the fact of one or
more such omissions. The table gives the derived estimates
of the precision and accuracy found for the single most
commonly used colorimetric method for each of the elements
in the sample. A limited amount of data for less commonly
used colorimetric and other methods showed a similar pattern.

One is struck immediately both by the low precision of
the results and the frequency of large absolute errors. The
data on the ranges of values reported reveal that it was not
an unusual event for a constituent to be missed entirely.
In view of the fact that averages of triplicates should be
more reliable than single values, and averages of 14 to 46
laboratories should be vastly more reliable than the observation of a single laboratory, one is forced to the conclusion that lacking supporting evidence, a single value
reported by one laboratory under the conditions of this
study carries very little weight indeed. This is a particularly disquieting thought in view of the fact that the
vast majority of existing data on trace element distribution
consists of single measurements by single laboratories
without benefit of any cross comparisons.

In Table 2 is given some results on the determination
of iron taken from a recent intercalibration study on trace
elements in sea water conducted by Brewer and Spencer at
the Woods Hole Oceanographic Institution (2). Thirteen

TABLE 2

INTERCALIBRATION STUDY: DETERMINATION OF
IRON (µg/Kg) IN TWO SAMPLES OF SEA WATER

| | SAMPLE A | | SAMPLE B | | RATIO |
METHOD	MEAN	SD%	MEAN	SD%	B/A
AAS	8.6	3	14.7	7	1.7
AAS	6.7	27	7.9	39	1.2
AAS	7.4	-	15.1	-	2.0
AAS	5.87	4	9.74	6	1.7
AAS	4.3	4	7.05	13	1.8
NAA	9.7	19	15.3	10	1.6
NAA	7.28	71	4.0	6	0.6
NAA	22	-	22	-	1.0
NAA	5.7*	9	5.1	-	0.9
COL	11.7	5	17.0	6	1.5
COL	7.67	3	13.33	4	1.7
PAA	28.3	4	45.0	4	1.6
NR	31.6	48	32.7	4	1.0
MEAN	12.2		16.6		1.4
MEAN**	7.7		12.5		1.6

 * Excludes one value
** Only replicated data with SD<20%

laboratories undertook to measure the iron content of three
subsamples taken from two large primary samples of sea water,
each laboratory using its own standard method. Five used
chelation-solvent extraction followed by atomic absorption
spectroscopy, four chose neutron activation analysis of
dried sea salts, two colorimetric procedures, one coprecipi-
tation with lanthanum hydroxide followed by atomic absorption, and one unreported. Again a striking lack of agreement
is observed both in the absolute values obtained and in the
precision of the data. The range of values reported was,
for both primary samples, 3.7 to 47 µg/kg, and the ratio of
the two sample concentrations ranged over more than a factor
of three. Similar patterns were obtained for the determinations of cobalt, copper, lead, manganese, nickel and zinc.

The discovery of a situation such as this must necessarily come as a shock to anyone accustomed to accepting
at face value typical published claims for the precision
and accuracy of trace analytical methods. Lest it be
assumed that the explanation is simply a particular inept-
ness on the part of water chemists, let me hasten to point
out that the same mournful lack of consistency has appeared
when geochemists have studied trace metals in standard rock
samples, when industries have cross-checked control laboratory performance; and when clinical laboratories have
collaborated in comparing unknown cholesterol samples. In
actual fact, behavior of this sort is to some degree
characteristic of all experimental measurements. Most
investigations are done in such a way that the discrepances
do not come to light, but when careful observations are
made--especially with demanding operations such as trace
metal analysis where the "signal to noise ratio" is
unfavorable--there is no overlooking the effect. The path
of the trace analyst is strewn with hidden pitfalls, to
say nothing of a generous number of more obvious potholes,
and it is profitable at this point to identify the more
important ones.

SOURCES OF VARIABILITY

First, it must always be kept in mind that the analysis
of real samples, the final result is the product of a multi-
stage process, each step of which is an input channel for
error. At the outset, one is faced with the problem of
sample definition--what is the sample we are getting? Is
it representative and/or homogeneous, and is it appropriate
to answer the chemical questions which prompt its taking?
Then there is the matter of physically taking the sample
itself, sometimes no mean feat. Normally there is a period
of storage, somewhere, in something, for some period of time.
The sample will be handled in the course of treatment and it
will often be measured in a quantitative subsampling step.
There is often chemical treatment of one sort or another
and there may be a separation and/or concentration process
involved. The determinative step, chemical or physical,
takes place and it must be coupled with some sort of
analog-to-digital conversion, be it nothing more complicated
than reading a buret. The information from a calibration
operation must be fed in, and the final step of the process
is the sometimes surprisingly subjective· expression and
interpretation of the result.

For evaluating the dependability of our results, we
need information on the precision and accuracy of the overall analytical process under actual conditions of use. All
too often the information in the literature is based on a
shamefully unrealistic approach. The proud author of a
new method is likely to have measured his reproducibility
under completely artificial conditions, if indeed he has
done more than demonstrate the repeatability of the determinative step on a known sample. With the best of intentions
he may do everything in his power to avoid encountering
the natural variability in his method, even to discarding
some of his observations because "obviously something
must have gone wrong with them." Data based on the repeatability of a method in the hands of a single person taking
identical portions of a clean standard solution to measure
all in one afternoon are irrelevant to the situation of the
environmental chemist. Under supposedly identical conditions, a method will often give significantly different
results on different days, at different times of the day
with different batches of reagents, with presumably
identical samples from different sources, and in the hands
of different people. This is simply a consequence of the
fact that the number of factors which affect a method or
a measurement is enormous. The major ones are usually
obvious, so we are aware of them and can make some effort
to control or avoid them. Some factors vary so little or
so rarely in our working environment that we overlook them,
or else the trouble goes away before we have the time to
figure out what is causing it. Many factors are actually
quite significant, yet we are unaware of their influence
simply because we have never set up a situation which
would reveal their effects. When we do, we are likely to
get results like those in the laboratory intercomparison
studies I have just described.

We are accustomed to the idea that instrumental measurements are susceptible to physical factors such as temperature,
but we tend to overlook or forget that there are so many such
factors. I have at one time or another run across measurements which were unexpectedly but noticably responsive to
magnetic fields, static charge, variations in line voltage
and frequency, the intensity and spectral distribution of
room lighting, humidity, the functioning of other not
necessarily nearby instruments, the pressure in the water
mains, vibration, the electrical conductivity of cooling
water, the cycling of the central air conditioning system

and, literally, the phase of the moon. Chemical reactions
have their own peculiarities many of which are even less
predictable than those of instruments. It should be
sufficient to point out that even the order in which the
starting materials are added to a reaction mixture may
affect not only the rate of formation and the purity but
even the identity of the principal product.

Personal equation is a well recognized component in
experimental differences. That some people have a knack
for making instruments or procedures work, while others
can't get the hang of it although both seem to be doing
just the same things, is a phenomenon familiar to all of
us. It emphasizes the fact that there are many unidentified
factors involved in our operation which we sometimes adjust
to quite unconsciously so we are able to do things
successfully although we cannot explain how. Another
important aspect of personal equation is the variation
between individuals in reading instruments and estimation
of final digits by interpolation. Gysel (3) in a study
of the unconscious number preferences of technicians
estimating the last significant figure in microchemical
weighings found that the statistically predicted equal
frequency of occurrence of the digits 0 to 9 was not even
roughly approached. Instead, each individual showed a
characteristic pattern of preferences in which the frequencies of appearance of some digits differed by factors as high
as 5 or 6. These characteristic individual patterns
sometimes showed remarkable stability and reproducibility
from year to year. Figure 1 shows the frequency distribution of a set of 1050 last digit estimations taken from the
laboratory notebooks of one of Gysel's technicians, and
is quite typical. The probability of obtaining such a
large deviation from the equal frequencies which should be
observed in drawing from a uniform population is less than
one in a million. Gysel found the same individual patterns
of number preference in weighing, reading verniers, colorimeter and flame photometer scales, and in estimating the
last digit in slide rule calculations. I have verified
the reality of the effect by examining the last digit in
buret readings recorded in student notebooks. Clearly
this phenomenon can be a serious source of bias when one
is straining to get the second figure in a two-significant
figure result, and seems to be present—no matter how
objective we think we are—in all of us.

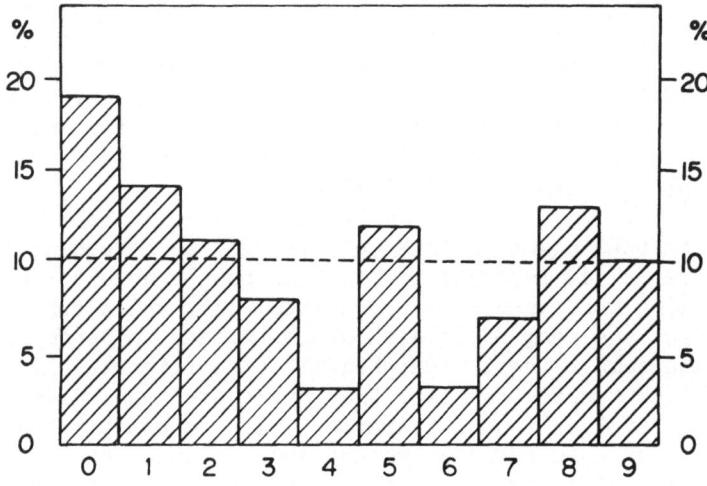

Figure 1. Frequency of occurrence of digits in the last (estimated) place in 1050 microchemical weighings by a single individual. [After Gysel (3)]

And that is another touchy point--none of us is as objective as he thinks he is. The knowledge of what a result ought to be or the advantage attached to having it come out a given way will bias the way we read it. When experiments are conducted to compare the results obtained under ordinary conditions with those obtained using "blind" samples, the open samples usually show us either yielding to temptation or bending over backwards to avoid it. Being human, we are likely to accept results which agree with our prejudices but repeat experiments which oppose them, instead of treating them all with impartiality.

PARTICULAR DIFFICULTIES

In addition to these general problems which, as I have said, apply to any quantitative scientific investigation, trace analysts must cope with four particular problems which are especially vexing. These are contamination, loss of constituent, matrix effects and sampling. Contamination of the sample by containers, apparatus,

reagents and the ambient air is sometimes extremely
difficult to avoid. Robertson (4) has discussed in considerable detail the role of contamination by materials
of construction, containers, solvents and reagents used
in the analysis of sea water. By neutron activation he
showed that many of the common materials which come in
contact with the sample contain high concentrations of
metallic impurities which could result in contamination.
Table 3 presents a small sample of his extensive data
comparing the trace metal content of sea water with that
of common materials with which it is likely to come in
contact. It should be kept in mind that while these figures
are fairly typical, any given batch of material may be
orders of magnitude higher or lower in concentration of
a contaminant than is shown here. Samples of quartz tubing
from six sources, for example, ranged from less than 0.01
up to 1940 ppb in antimony content.

Concealed sources of contamination are an especial
nuisance. Teflon, as manufactured, is usually very low in
trace metals, but in the process of fabrication into
chemical apparatus particles of grit, rust and dirt may
become imbedded in the surface to act as a source of
contamination for long periods of time. Many workers,
after going to great effort to procure zinc-free bottles
or vials for liquid samples have discovered that the rubber
or plastic gaskets on the caps have been compounded with
several percent of zinc oxide. The performance of a
platinum hypodermic needle is distinctly disappointing if
it has been attached to its hub with a conventional brazing
alloy containing about 50% copper.

Laboratory reagents may be highly contaminated with
both metals and non metals. Natural sea water, for example,
contains much lower concentrations of many heavy metals
than artificial sea water made up from the purest available
reagent grade chemicals. Water for laboratory use presents
a dilemma: ion exchange removes most metallic impurities
well but introduces organic matter, while distillation
reduces the level of organic contamination but tends to
introduce metals. The ambient air can be a major source
of metal-bearing particulates and some workers find the
use of a positive-pressure, laminar-flow, clean hood
mandatory.

The loss of trace metals from samples in storing and

TABLE 3

TRACE ELEMENT CONTENT OF SEA WATER AND
COMMON LABORATORY MATERIALS: P.P.B.

	Zn	Fe	Sb	Cr
Sea Water	10	10	0.3	0.5
Pyrex	730	2.8×10^5	2.9×10^3	--
Quartz	25	--	40	200
Polyethylene	25	1.1×10^4	0.8	19
Teflon	9	35	0.4	<30
Rubber	4.1×10^7	<100	360	4.2×10^5
PVC	7.1×10^3	2.7×10^5	2.7×10^3	2
Plexiglas	<10	<140	<0.01	<10
Millipore	2.4×10^3	330	39	1.8×10^4
Kimwipe	4.9×10^4	1×10^3	16	500

handling is as great or greater a problem. The processes
by which losses may take place include adsorption, precipitation, permeation, ion exchange, volatilization, liquid-
liquid extraction, and sequestration by chemical or biological action. It should be noted that this list includes
nearly all of the commonly used analytical separation
process. However difficult it may be for the analyst to
make them function properly for a desired purpose, nature
seems to have no problem getting them to perform effectively
for undesirable purposes.

Adsorption is a widespread phenomenon wherever
extensive surface area is exposed to samples. In the
analysis of water, suspended particles of sediments, fly
ash, soot, organisms and detritus can remove significant
fractions of the dissolved trace substances. Container
walls of glass or other silicious materials are very
effective in holding trace metal ions which polymerize on
hydrolysis, lead being a prominent example. Glass is also
an ion exchanger, as are plastics which have acquired ionic
functional groups by treatment with strong nitric or
sulfuric acid. The formation of a precipitate may carry
down trace metals from solution either by adsorption or by
occlusion within the crystal lattice of the precipitate.
The concentration factors are sometimes very large.

It is quite surprising how many metallic elements may
be lost as a consequence of the volatility of their compounds.
Conditions which favor the formation of covalent halides,
e.g. dehydrating media such as hot, fairly concentrated
sulfuric, perchloric, or phosphoric acid in the presence of
halide ions have been shown by Hoffman and Lundell (5) to
permit the quantitative distillation of chromium, arsenic,
antimony, tin, selenium, rhenium, ruthenium and osmium, and
the vaporization of substantial amounts of germanium,
molybdenum and mercury. The usefulness of the graphite
furnace technique in the flameless atomic absorption
technique for the determination of metals such as lead and
cadmium emphasizes the potential hazards in traditional
dry ashing procedures quite apart from the danger of loss
by incorporation in particulate emissions. The ready volatility of mercury, both as methyl mercury and as the element
has received widespread attention recently, and a number of
procedures for the determination of trace mercury as based
on its reduction to the element and volatilization from
solution at room temperature. It has been observed (6)
that microorganisms normally present in the laboratory are
capable of rapidly volatilizing mercury as the metal from
plasma, broth or urine samples. The paper by Toribara,
Shields and Koval (7) on radioactive exchange in dilute
solutions of mercury compounds gives a revealing and enter-
taining account of the effectiveness of a little mercury
vapor in the system. The possibility of a highly diffusable
component permeating solid or gel phases must also be kept
in mind. Mercury vapor readily passes into (and through)
rubber membranes and the walls of warmed polyethylene
containers.

The extraction of metal ions as chelate complexes,
based on their preferential solubility in water-immiscible
organic liquids, is a widely used analytical procedure. It
is not generally recognized, however, that many environmental water samples contain naturally occurring chelating
agents derived from the decomposition of plant and animal
matter, and from the metabolic byproducts of microorganisms.
The resulting metal chelate complexes may then partition
preferentially into the high molecular weight organic
compounds which make up the greasy film that forms on the
surface of glassware in contact with dirty water, or into
the layer of water-insoluble plasticizer on the surface of
an unwisely chosen plastic bottle. Chelating agents which
are surface active may concentrate trace metals at interfaces,

such as a liquid-liquid interface (often discarded in
extraction procedures) or in foam.

A frequently overlooked mechanism for the loss or
inactivation of trace components is biological action. Algal
or bacterial growth may take place with surprising rapidity
in stored water samples, and concentration factors may be
large. The effects are most striking with nutrient factors
such as phosphorus and nitrogen—the use of unsterilized
plastic bottles has been known to result in loss of most of
the phosphate from water samples through incorporation in
bacterial colonies dwelling on the bottle walls—but may
also be significant for metals.

The term "matrix effects" is used to cover a miscellaneous collection of phenomena, sometimes explainable
sometimes not, but associated with the presence of other
components in the sample. A typical example is the tendency
of chelating or other strongly complexing ligands to inhibit
the characteristic reactions of metal ions to such an extent
that normal analytical methods produce low results or miss
the component entirely. A very strong or kinetically
hindered complex with a naturally occurring chelating agent
in the sample may prevent the formation of an extractable
complex with a chelating reagent, inhibit a color forming
reaction, or impede the reduction of an ion at an electrode.
The phenomenon of water with a negative hardness value has
surprised more than one investigator. Ordinary chemical
interferences when unsuspected sometimes appear as matrix
effects. A notable example was pointed out by Chamberlain
and Shapiro (8) who observed that the commonly used colorimetric method for phosphate in potable water also responded
to arsenate, and as a consequence certain Minnesota lakes
which had been thought to contain considerable amounts of
phosphate were actually "clean" in that respect, although
they contained rather large but unsuspected quantities of
arsenate. Also worthy of mention is the as yet unexplained
discrepancy between the results from the chelation-extraction-
atomic absorption method for determining cobalt in sea water
and the results obtained by neutron activation analysis of
the freeze-dried salts. Brewer and Spencer (2) pointed out
that the extraction procedure would be expected, if anything,
to be low because of incomplete recovery—a matrix effect
caused by natural chelating agents—while the activation
method should give a good measure of total metal. On
comparing the results reported by eight laboratories on two

samples, four using neutron activation and four using
extraction-atomic absorption, it was found that the averages by the extraction procedures were five to six fold
larger than those by activation.

Sampling is frequently a weak link in the chain of
analytical operations. Unglamorous and often inconvenient,
although essential to the result, the sampling operation
may be given far less thought and care than it deserves.
It is necessary first to define the system being measured,
and then to establish what constitutes a representative
sample of it. If more than one phase is present, either
separation of the phases or proportional representation of
them is needed. A common source of perplexity is the
presence of particulate matter in water or other fluids.
Are the particles to be excluded or included? If excluded,
what is to be taken as the definition of a particle? The
biologists convention that anything in water which passes
through a 0.45 micron Millipore filter is regarded as
dissolved is not altogether satisfactory to a chemist. The
changes in the properties of natural systems before, during
and after sampling are exceedingly difficult to allow for.
In the study of environmental systems, their dynamic nature
may call for a prohibitively large number of measurements
to give a clear picture of their behavior. What samples,
in what number and with what frequency are necessary? If
continuous sampling is to be done, how is it to be made
representative?

CONCLUSIONS

I hope that as a result of this discussion, I have made
clear that environmental trace metal analysis is indeed well
supplied with pitfalls: big, deep ones with sharpened
bamboo stakes in the bottom. It is not my intention,
however, to suggest that such analysis cannot be done, and
done well. The alert and informed investigator who has
the patience and the support to study and refine each step
of the long and complex process leading from sample taking
to final result, who looks at the overall performance of
his method as well as its individual parts, and who subjects
it to rigorous, objective evaluation by intercomparison with
other laboratories on blind samples has a good chance of
getting truly meaningful results. It must be recognized
that this is a tedious and expensive process, but by-passing

it is no economy in the long run. What is much needed now
is both a thorough development and testing of the methods
used, and an adequate revelation of the findings in publications. When this is done, we should have both a sound
basis for judging the validity of results which are
reported, and a measure of protection against embarrassing
surprises.

REFERENCES

(1) R. J. Lishka, Water Metals No. 3. U. S. Department of
    Health, Education and Welfare, Public Health Service,
    Cincinnati, Ohio (1965).
(2) P. G. Brewer and D. W. Spencer, Trace Element Inter-
    calibration Study Reference No. 70-62, Woods Hole
    Oceanographic Institution, Woods Hole, Mass. Dec.
    1970 (Unpublished Manuscript).
(3) H. Gysel, Mikrochim. Acta 1953 266; 1956 577.
(4) D. E. Robertson, Anal. Chem. 40 1067 (1968).
(5) J. I. Hoffman and G. E. F. Lundell, J. Research National
    Bureau of Standards 22 465 (1939).
(6) L. Magos, A. A. Tuffery and T. W. Clarkson, Brit. J.
    Industr. Med. 21 294 (1964).
(7) T. Y. Toribara, C. P. Shields and L. Koval, Talanta 17
    1025 (1970).
(8) W. Chamberlain, J. Shapiro, Limnol. Oceanog. 14 921
    (1969).

Acknowledgement: This work was supported in part by the
United States Atomic Energy Commission under Contract
AT(11-1) 2169.

Herbert L. Volchok
Donald C. Bogen
Health and Safety Laboratory
U. S. Atomic Energy Commission
New York, New York 10014

ABSTRACT

Two programs of trace metal research are being carried out at the Health and Safety Laboratory (HASL). (a) In the surface air program, continuous sampling of the particulates in surface air, has been performed since mid 1967, at about 20 stations extending over almost the entire Western Hemisphere. The samples are analyzed for Pb, on a monthly basis. New York City generally has the highest concentration, of these sites, averaging about 2 $\mu g/m^3$; Miami and San Juan, Puerto Rico are usually within a factor of two of New York. In the Southern Hemisphere, Santiago Chile, often reaches 1 $\mu g/m^3$ and exhibits a pronounced winter peak in lead concentration.

(b) In the fallout program, V, Cr, Mn, Ni, Cu, Zn and Pb have been analyzed in monthly total and "wet" samples collected in New York City. In general the levels of trace metals at this site approach or exceed any U. S. values reported. The seasonal variations and relationship of wet to total fallout, will be discussed.

INTRODUCTION

One of the most important and perhaps most painful lessons to be learned from the environmental contamination and pollution condition in which we find ourselves today,

is that we must improve our foresight. We simply cannot
afford, many more times, to start making observations and
deciding on courses, after damage has been detected. The
costs are too high and in some cases the effects may be
irreversible. To counteract this trend, data must be collected and collated with regard to many environmental
additives. Then, should a deleterious effect appear,
pertinent information on sources and concentrations would
be available, and prompt corrective action would at least
be possible.

The Health and Safety Laboratory (HASL) operates a
number of programs of systematic sampling and analysis of
radionuclides on national and international scales. Some
examples are: (a) worldwide fallout collections[1], (b)
Western Hemisphere surface air sampling[2], (c) stratospheric
air sampling[3,4], and (d) studies of radionuclides in U. S.
diets[5]. These, of course, are all related primarily to
the radioactive contamination of people and the earth, by
the debris from nuclear explosions. However, they serve as
good examples of remote sampling networks and, involving co-
operation of individuals, government agencies and private
institutions in many parts of the world demonstrate the
diplomatic and logistical problems associated with programs
of this type. Our experience has proven that these systems
can work effectively and that valuable samples and data are
obtainable.

The Health and Safety Laboratory has now become interested in obtaining other, non-nuclear, basic data on
potentially harmful environmental contaminants. We believe
that we can apply the experience gained in the fallout
programs, the sample collection systems, the establishment
of networks, and the handling and dissemination of data, to
other similar global problems.

In this paper we describe HASL's excursion into the
realm of trace metals. Our motive is primarily to provide
systematic and accurate "base line" data, hoping that these
will, in time aid in understanding, treating and maybe even
avoiding future problems in environmental pollution.

SAMPLING

To date, we have been studying two types of samples,
(a) the particulates in surface air, and (b) fallout.

Surface Air Samples

This trace metal study is an outgrowth of the HASL
Surface Air Sampling Program. In this, continuous air
sampling has been taking place since early 1963, at about 20
stations in the Western Hemisphere mainly for studies of
concentrations and distributions of radionuclides in ground
level air[2]. The primary orientation of the network is
approximately along the 80th meridian (west) following the
east coast of North America and the west coast of South
America. A few special purpose sites were also set up; for
instance at high elevations, Mauna Loa Hawaii, at about 3400
meters and Chacaltaya Bolivia, at about 5200 meters; and
island stations at Bimini, Bahamas and Easter Island, west
of Chile. In addition, for short periods we studied lati-
tudinal effects and also compared concentrations over the
sea vs. over the land[6]. The current network extends from
almost 82 degrees north latitude, at Nord, Greenland to the
South Pole. The current sampling sites are listed in Table
1, along with pertinent geographical data, and mapped on
Figure 1.

The samples are collected by a Roots-Connersville
blower, at the rate of about 1500 cubic meters of air per
day, through a 20 cm diameter Microsorban filter. With
minimum care and maintenance most of these units have
operated trouble free and on a continuous basis for almost
eight years.

Lead analysis on surface air filters commenced in mid
1967, except for a few months of data in 1966. Our purpose
in this is to develop a set of "base line" data for lead
concentrations in air, as a function of time, and over a
variety of geographical areas. These are not meant to
represent background; but rather a series of examples of
lead concentrations in air over a wide spectrum of sites
ranging from the densely populated New York City and
Santiago, Chile areas to the South Pole; and from sea level
to over 5000 meters on the top of a mountain.

TABLE 1. HASL SURFACE AIR SAMPLING STATIONS

Stations	Latitude	Longitude	Elevation (m)
*Nord, Greenland	81°40'N	17°00'W	250
*Thule, Greenland	76°36'N	68°35'W	259
Bravo Ocean Station	56°30'N	51°00'W	0
Charlie Ocean Station	52°45'N	35°30'W	0
*Moosonee, Ontario, Canada	51°16'N	80°30'W	10
Delta Ocean Station	44°00'N	41°00'W	0
*New York, New York	40°48'N	73°58'W	38
*Salt Lake City, Utah	40°46'N	110°49'W	1516
*Sterling, Virginia	38°58'N	77°25'W	76
Echo Ocean Station	35°00'N	48°00'W	0
*Miami, Florida	25°49'N	80°17'W	4
*Bimini, Bahamas	25°46'N	79°22'W	3
*Mauna Loa, Hawaii	19°28'N	155°36'W	3401
*San Juan, Puerto Rico	18°26'N	66°00'W	10
*Balboa, Panama Canal Zone	8°58'N	79°34'W	23
*Guayaquil, Ecuador	2°10'S	79°52'W	7
*Lima, Peru	12°03'S	77°08'W	13
*Chacaltaya, Bolivia	16°21'S	68°07'W	5220
*Antofagasta, Chile	23°37'S	70°16'W	31
*Isla de Pasqua (Easter Is.)	27°10'S	109°26'W	41
Portillo, Chile	32°50'S	70°08'W	2850
*Santiago, Chile	33°27'S	70°42'W	520
*Puerto Montt, Chile	41°27'S	72°57'W	7
*Punta Arenas, Chile	53°08'S	70°53'W	35
*Gabrial Gonzalez Videla, Chile	64°49'S	62°52'W	10
*South Pole Station	90°00'S	–	2800

*Stations currently being sampled.

Figure 1. Surface Air Program sampling sites.

Fallout Samples

Our studies in this program, to date, have been confined to New York City. In this, our original objective was to determine fallout levels of eight trace metals: vanadium, chromium, manganese, nickel, copper, zinc, lead and cadmium. Sampling was started in late 1969 on the roof of the HASL building, 13 floors above street level, in the southwestern area of Manhattan. Initially we collected only total fallout, that is, everything that falls in and is retained in a steep-walled stainless steel pot, about 12 inches in diameter and about 14 inches tall. This pot collector has been in use for many years in our radioactive fallout studies and has proven to be the most reliable collector of any tested.

A second total fallout collection was initiated in March of 1971 when samplers were placed on the roof of the RCA building, 72 stories above street level, in mid Manhattan. With these data we hope to learn something about the homogeneity of the lower troposphere with respect to these particular sources.

In September, 1970, we began collections of "wet" fallout, in addition to the total. These are samples restricted solely to periods of precipitation. The original equipment for this program was loaned by the Air Pollution Control Office (APCO), (it had been commercially available, but is no longer). When moisture falls on the printed circuit board sensor, a current flows and the cover of the housing opens exposing a plastic collector of about 15 cm diameter. When precipitation ceases, a heater dries the sensor, breaking the circuit and causing the device to close.

This system proved to be inadequate for the dirt and weather conditions in New York City. The mechanical system for moving the cover tended to clog and freeze, and the collector was too unreliable for unattended periods of more than just a few days.

HASL's own Instrumentation Division has since designed a new wet/dry fallout collector, now undergoing tests on the roof. In this instrument, two standard fallout pots are used, one exposed during dry periods and the other during precipitation. The sensor, mounted on the cover works on same principal as was described for the previous instrument.

ANALYSIS AND QUALITY CONTROL

Surface Air Samples

The surface air samples have been analyzed by commercial contractor laboratories since the onset of the program at HASL. With specific regard to the lead data, a number of organizations have participated. In the first half of 1967, as part of the contractor selection procedure, three laboratories each contributed two months of data, as follows:

> Jan. and Feb. 1967 - Tracerlab Inc., Richmond, Cal.
> Mar. and Apr. 1967 - Custom Nuclear, Stanford, Cal.
> May and June 1967 - Isotopes Inc., Palo Alto, Cal.

From July 1967 to the present the analyses have all been performed by Trapelo Division/West of Richmond, California (formerly called Tracerlab).

The analytical procedure presently utilized by Trapelo Division/West consists of dry ashing of the microsorban filter paper at 420°C for 24 hours. The ash is dissolved in nitric acid and lead is precipitated as hydroxide, dissolved in HCl and passed through a Dowex 1x8 100-200 mesh ion exchange column. The eluate is then evaporated to a small volume and analyzed on a Perkin-Elmer 303 Atomic Absorption spectrophotometer.

Quality of the Pb analysis is monitored by submission of blanks and knowns, camouflaged as samples, with every sample shipment.

The blank has proven variable, but has remained always less than about 200 μg per sample, and for the most part, in 1970 averaged about 50 μg per sample. In view of the normal one month sampling times (about 40,000 m^3) and the levels of lead concentration encountered, this blank is sufficiently below environmental levels, as to be virtually negligible.

The accuracy of the Pb analysis in terms of the percent deviation of the measured values from the values expected, has averaged slightly less than 20%, over the period of study.

Periodic reports of the quality of analyses in the HASL Surface Air Program are published[7].

Fallout Samples

The fallout samples are all analyzed at HASL by the Radiochemistry Division.

At the end of each monthly sampling period, the collectors are rinsed repeatedly with 1:1 nitric acid and distilled water to quantitatively remove all of the solid material for analysis.

The sample is then reacted three times with aqua regia, evaporating each time to a volume of about 50 ml. After the final evaporation, the solution is diluted with an equal volume of distilled water, and filtered on a 0.22 micron Millipore filter. The filtered residue is again reacted three times with aqua regia, filtered as before, and both filtrates are combined, evaporated and diluted to a known volume and to a concentration of 1 $\underline{N}$ HNO_3.

The resulting solutions are analyzed on a Perkin-Elmer 303 Atomic Absorption Spectrophotometer. Table 2 lists the conditions for the trace metals analyzed in this program.

TABLE 2. CONDITIONS FOR ATOMIC ABSORPTION OF TRACE METALS

Element	Wavelength (A^o)	Flame Type	Sensitivity $(\mu g/ml/1\%)$	Working Range (PPM)
Cadmium	2288	Air - Acetylene	0.04	0.5 - 5.0
Chromium	3579	N_2O - Acetylene	0.10	2 - 20
Copper	3247	Air - Acetylene	0.15	2 - 20
Manganese	2795	Air - Acetylene	0.10	2 - 20
Nickel	2320	Air - Acetylene	0.15	2 - 25
Lead	2837	Air - Acetylene	0.5	10 - 100
Vanadium	3183	N_2O - Acetylene	1.7	5 - 100
Zinc	2138	Air - Acetylene	0.025	0.2 - 3.0

Standards for the instrument calibration were obtained commercially and prepared by dilution to the appropriate volume. Sodium was added to all standards to suppress the formation of ionic species in the flame; this is especially important in the measurement of vanadium. Calibration curves were run concurrently with the sample measurement to compensate for instrument variation.

Recoveries were evaluated by conducting experimental runs with known amounts of trace metals. The data for these tests shown in Table 3 indicate that in general the recoveries are 95% or greater.

TABLE 3.   RECOVERY IN TRACE METALS ANALYSIS

Run	Cd	Cr	Cu	Mn	Ni	Pb	V	Zn
				% Recovery				
A	97.5	100	100	102	105	95.0	92.5	101
B	97.5	97.5	100	102	105	94.0	92.5	101
C	94.0	97.5	100	100	105	100	95.0	99.0
D	95.0	100	100	100	105	94.0	95.0	99.0

Reagent and apparatus blanks are periodically run; the results of eight such tests are listed in Table 4.

TABLE 4.   REAGENT BLANKS

Run	Cd	Cr	Cu	Mn	Ni	Pb	V	Zn
				μg/sample				
Blank A	0.5	6	4	1	3	4	N.D.	112
Blank B	1	2	18	1	5	6	N.D.	132
Blank C	1	4	12	1	2	3	N.D.	103
Blank D	1	5	6	1	6	7	N.D.	145
Blank E	0.5	3	9	1	4	4	N.D.	152
Blank F	0.5	4	11	1	3	6	N.D.	127
Blank G	1	3	7	1	5	8	N.D.	134
Blank H	1	5	14	1	2	3	N.D.	116

N.D. = Not detectable.

The efficiency of the leaching procedure was evaluated by re-leaching the residues of seven samples, as described in the procedure above. The final filtrates were analyzed for trace metals. The results listed in Table 5 show that in general, except for low concentration samples, the residue contain negligible fractions of the original metal content.

Precision of the analytical procedure is monitored by periodic analysis of a "standard" aliquot of fallout material, submitted to the analysts as an unknown. The results of 5 such tests, shown in Table 6 along with the means and standard deviations, indicate that the reproducibility of analysis for these trace metals in fallout is 10% (1σ) or less.

TABLE 5. EFFICIENCY OF THE LEACHING PROCEDURE

Sample No.	Cd	Cr	Cu	Mn	Ni	Pb	V	Zn
				μg/sample				
P5126 (Initial Value)	80	5560	2500	1950	4400	7250	6690	8750
P5126 (Residue Value)	1.3	73	50	15	25	56	63	62
% Residue	1.6	1.3	2.0	0.8	0.6	0.8	0.9	0.7
P5129 (Initial Value)	110	6125	3700	2450	5150	6750	7500	10625
P5129 (Residue Value)	1.3	185	14	6	15	13	13	45~
% Residue	1.2	3.0	0.4	0.3	0.3	0.2	0.2	0.4
P5162 (Initial Value)	120	5190	3125	2900	5300	13750	8563	10940
P5162 (Residue Value)	1.3	80	15	35	N.M.	40	63	66
% Residue	1.2	1.5	0.5	1.2	-	0.3	0.8	0.6
P5338 (Initial Value)	17	650	825	500	400	2500	718	2030
P5338 (Residue Value)	0.25	35	N.M.	5	7.5	2.5	N.D.	36
% Residue	1.5	5.4	-	1.0	1.9	0.1	-	1.8
P5369 (Initial Value)	75	4000	2450	1900	2825	9250	1825	7190
P5369 (Residue Value)	3.5	43	13	13	23	13	13	47
% Residue	4.7	1.1	0.5	0.7	0.8	0.1	0.7	0.6
W0457 (Initial Value)	20	N.D.	30	2.5	185	50	N.D.	450
W0457 (Residue Value)	1.0	N.D.	7.5	N.D.	7.5	2.5	N.D.	41
% Residue	5.0	-	25	-	4.1	5.0	-	9.0
W0458 (Initial Value)	30	17	50	7.5	75	110	13	194
W0458 (Residue Value	0.25	N.D.	10	N.D.	N.D.	2.5	N.D.	22
% Residue	0.8	-	20	-	-	2.3	-	11

N.D. = Not detectable
N.M. = Not measured

TABLE 6. PRECISION OF ANALYSIS

	Cd	Cr	Cu	Mn	Ni	Pb	V	Zn
				μg/sample				
	105	17000	3900	3250	14000	7100	12000	10000
	105	18000	3800	3600	15000	7250	12000	10000
	105	20500	3900	3250	16750	7500	12000	9650
	105	22000	3900	3400	16750	7500	12000	10000
	116	18500	2750	3100	14250	8000	13500	8800
Avg.	107	19200	3850	3300	15350	7500	12300	9700
% S.D.	4.7	10	1.8	5.8	8.8	4.0	5.8	5.3

Table 7

STABLE LEAD CONCENTRATIONS IN SURFACE AIR DURING 1966
(μg/m³) x 100

SITE	JAN.	FEB.	MAR.	APR.	MAY	JUNE	JULY	AUG.	SEP.	OCT.	NOV.	DEC.
THULE, GREENLAND	--	--	--	--	--	--	--	--	--	0.76	--	--
BRAVO OCEAN STATION	--	--	--	--	--	--	--	--	--	0.61	0.92	4.38
CHARLIE OCEAN STATION	--	--	--	--	--	--	--	--	--	0.42	2.21	8.36
MOOSONEE, ONTARIO	--	--	--	--	--	--	--	--	--	3.52	4.75	7.12
SEATTLE, WASHINGTON	--	--	--	--	--	--	--	--	--	37.20	197.00	195.00
DELTA OCEAN STATION	--	--	--	--	--	--	--	--	--	0.11	1.24	5.15
NEW YORK, NEW YORK	--	--	--	--	--	--	--	--	--	209.00	298.00	216.00
STERLING, VIRGINIA	--	--	--	--	--	--	--	--	--	38.40	44.60	79.90
ECHO OCEAN STATION	--	--	--	--	--	--	--	--	--	0.73	0.92	43.60
MIAMI, FLORIDA	--	--	--	--	--	--	--	--	--	88.80	137.00	201.00
BIMINI, BAHAMAS	--	--	--	--	--	--	--	--	--	3.24	31.90	4.00
MAUNA LOA, HAWAII	--	--	--	--	--	--	--	--	--	0.84	--	9.81
SAN JUAN, PUERTO RICO	--	--	--	--	--	--	--	--	--	47.30	99.00	57.30
BALBOA, PANAMA	--	--	--	--	--	--	--	--	--	16.90	18.00	18.40
GUAYAQUIL, ECUADOR	--	--	--	--	--	--	--	--	--	13.90	36.60	31.80
LIMA, PERU	--	--	--	--	--	--	--	--	--	20.80	99.30	47.40
CHACALTAYA, BOLIVIA	--	--	--	--	--	--	--	--	--	1.11	69.30	18.90
ANTOFAGASTA, CHILE	--	--	--	--	--	--	--	--	--	0.17	8.58	71.90
ISLE DE PASQUA-EASTER IS	--	--	--	--	--	--	--	--	--	--	2.50	--
PORTILLO, CHILE	--	--	--	--	--	--	--	--	--	--	--	1.52A
SANTIAGO, CHILE	--	--	--	--	--	--	--	--	--	25.10	49.00	19.70
PUERTO MONTT, CHILE	--	--	--	--	--	--	--	--	--	--	--	2.33
PUNTA ARENAS, CHILE	--	--	--	--	--	--	--	--	--	1.36	2.92	5.00

-- NO DATA

Table 7 (Cont'd)

STABLE LEAD CONCENTRATIONS IN SURFACE AIR DURING 1967
(μg/m^3) × 100

SITE	JAN.	FEB.	MAR.	APR.	MAY	JUNE	JULY	ALG.	SEP.	CCT.	NCV.	CEC.
THULE, GREENLAND	--	--	--	--	--	--	--	1.00	1.00	--	0.46	1.13
BRAVO OCEAN STATION	--	--	--	--	--	--	--	--	1.00	3.00	1.73	1.43
CHARLIE OCEAN STATION	--	--	--	--	--	--	--	--	1.00	--	2.75	0.31
MOOSONEE, ONTARIO	--	--	--	--	--	--	--	5.00	11.50	5.00	11.50	9.65
SEATTLE, WASHINGTON	--	160.00	--	--	--	--	82.00	--	--	--	--	--
DELTA OCEAN STATION	--	--	--	--	--	--	--	--	4.00	1.99	0.40	0.48
NEW YORK, NEW YORK	--	--	--	--	--	--	302.00	277.00	--	343.00	336.00	143.00
STERLING, VIRGINIA	--	--	--	--	--	--	--	52.00	55.00	86.00	109.00	57.50
ECHO OCEAN STATION	--	--	--	--	--	--	--	--	15.00	2.54	4.31	2.87
MIAMI, FLORIDA	--	--	--	--	--	--	--	195.00	214.00	279.00	272.00	245.00
BIMINI, BAHAMAS	--	--	--	--	--	--	5.00	5.00	4.92	6.00	10.20	7.58
MAUNA LOA, HAWAII	--	--	--	--	--	--	--	--	--	1.00	3.76	0.60
SAN JUAN, PUERTO RICO	--	--	--	--	--	--	--	69.00	110.00	141.00	141.00	88.60
BALBOA, PANAMA	--	--	--	--	--	--	17.00	20.00	13.00	31.00	25.80	29.20
GUAYAQUIL, ECUADOR	--	--	--	--	--	--	23.00	27.00	25.00	43.00	61.70	39.20
LIMA, PERU	--	--	--	--	--	--	38.00	34.00	45.00	29.00	40.40	48.20
CHACALTAYA, BOLIVIA	--	--	--	--	--	--	1.00	1.00	--	2.00	3.40	1.23
ANTOFAGASTA, CHILE	--	--	--	--	--	--	5.00	9.00	6.00	4.00	9.65	6.97
ISLE DE PASCUA-EASTER IS	2.50	2.86	2.90	2.76	--	--	--	2.22	--	--	--	--
SANTIAGO, CHILE	--	--	--	--	--	--	112.00	133.00	77.00	76.00	58.80	69.30
PUERTO MONTT, CHILE	--	--	--	--	--	--	--	--	--	--	0.25	0.47
PUNTA ARENAS, CHILE	--	--	--	--	--	--	8.00	9.00	7.00	8.00	6.79	6.99
PEDRO AGUIRRE CERDA	--	--	--	--	--	--	--	--	--	--	0.35	--

-- NO DATA

Table 7 (Cont'd)

STABLE LEAD CONCENTRATIONS IN SURFACE AIR DURING 1968
$(\mu g/m^3) \times 100$

SITE	JAN.	FEB.	MAR.	APR.	MAY	JUNE	JULY	AUG.	SEP.	OCT.	NOV.	DEC.
THULE, GREENLAND	0.21	0.72	0.21	--	1.11	--	--	1.82	2.97	1.00	1.79	1.11
BRAVO OCEAN STATION	1.79	3.52	0.16	1.71	1.09	1.35	4.51	21.20	4.51	1.28	--	--
CHARLIE OCEAN STATION	0.47	1.06	2.46	1.27	1.41	5.88	2.77	--	2.16	1.87	1.95	0.53
MOOSONEE, ONTARIO	7.00	2.22	5.83	2.60	4.24	2.66	3.59	6.15	9.56	--	3.35	4.67
DELTA OCEAN STATION	7.65	--	23.20	4.91	0.88	1.77	3.35	12.40	2.45	1.82	--	0.96
NEW YORK, NEW YORK	164.00	156.00	340.00	225.00	319.00	273.00	143.00	149.00	104.00	134.00	132.00	302.00
STERLING, VIRGINIA	293.00	63.70	65.30	21.20	43.50	53.10	67.50	66.80	57.20	56.60	57.40	45.80
ECHO OCEAN STATION	2.21	--	12.10	17.40	10.20	2.11	12.80	10.30	22.80	16.80	2.36	2.24
MIAMI, FLORIDA	104.00	200.00	162.00	145.00	133.00	131.00	115.00	201.00	204.00	192.00	190.00	183.00
BIMINI, BAHAMAS	4.66	11.80	15.20	12.90	11.20	35.00	11.00	16.30	5.36	17.40	--	--
MAUNA LOA, HAWAII	0.30	0.32	0.31	--	--	--	--	0.18	1.35	1.74	--	--
SAN JUAN, PUERTO RICO	95.90	168.00	44.00	33.40	67.70	81.20	71.80	95.70	152.00	147.00	138.00	78.70
BALBOA, PANAMA	21.90	37.50	30.20	15.90	29.60	29.20	18.40	29.90	22.90	29.60	24.20	25.70
GUAYAQUIL, ECUADOR	49.30	50.90	55.20	21.70	41.80	35.80	23.90	23.90	27.00	28.60	23.00	27.00
LIMA, PERU	52.40	75.10	52.50	32.00	53.80	40.40	43.70	49.30	55.40	32.30	43.30	36.30
CHACALTAYA, BOLIVIA	0.40	4.74	0.32	--	3.02	2.09	0.96	2.26	2.86	3.56	1.54	--
ANTOFAGASTA, CHILE	11.60	8.83	10.00	4.73	8.01	7.23	5.90	6.81	6.06	7.16	4.74	5.61
PORTILLO, CHILE	--	--	--	--	--	--	2.35	1.21	2.52	--	--	--
SANTIAGO, CHILE	71.40	49.10	95.60	194.00	138.00	226.00	106.00	84.10	61.10	27.90	49.50	16.30
PUERTO MONTT, CHILE	0.24	0.27	0.32	--	--	--	--	--	0.76	0.75	--	--
PUNTA ARENAS, CHILE	6.23	5.91	7.82	2.45	7.13	5.04	7.37	6.01	7.02	6.78	3.96	--
GABRIEL GONZALEZ VIDELA	--	--	--	--	--	--	--	--	--	--	--	0.62

-- NO DATA

Table 7 (Cont'd)

STABLE LEAD CONCENTRATIONS IN SURFACE AIR DURING 1969
$(\mu g/m^3) \times 100$

SITE	JAN.	FEB.	MAR.	APR.	MAY	JUNE	JULY	AUG.	SEP.	OCT.	NOV.	DEC.
THULE, GREENLAND	--	0.90	1.12	2.18	4.16	1.11	1.07	1.73	0.72	0.09	--	--
BRAVO OCEAN STATION	--	2.98	1.11	--	--	--	--	--	--	--	--	--
CHARLIE OCEAN STATION	1.59	--	2.42	--	--	--	--	--	--	--	--	--
MOOSONEE, ONTARIO	6.54	6.00	6.50	10.30	13.30	9.61	25.80	24.70	13.80	9.90	6.60	6.02
DELTA OCEAN STATION	--	6.00	41.70	--	--	--	--	--	--	--	--	--
NEW YORK, NEW YORK	99.70	33.60	153.00	242.00	241.00	218.00	--	38.70	159.00	245.00	153.00	116.00
STERLING, VIRGINIA	111.00	--	56.60	38.30	42.90	60.20	31.10	22.00	45.70	41.40	44.10	120.00
ECHO OCEAN STATION	--	21.00	1.25	--	--	--	--	--	--	--	--	--
MIAMI, FLORIDA	191.00	--	193.00	135.00	167.00	19.40	165.00	176.00	152.00	163.00	211.00	269.00
BIMINI, BAHAMAS	6.63	11.00	6.30	5.99	4.88	4.37	4.60	3.08	3.41	4.21	--	7.69
MAUNA LOA, HAWAII	--	0.80	--	--	--	1.13	0.32	0.34	1.42	0.07	--	--
SAN JUAN, PUERTO RICO	78.30	164.00	107.00	171.00	182.00	92.90	135.00	99.30	185.00	203.00	156.00	157.00
BALBOA, PANAMA	22.50	24.10	24.10	29.10	31.70	35.30	15.10	29.30	22.50	36.70	22.60	18.20
GUAYAQUIL, ECUADOR	29.20	31.50	31.40	42.70	33.20	27.40	16.60	14.10	21.30	33.20	40.30	40.00
LIMA, PERU	5.20	47.50	52.50	65.90	20.50	45.10	3.06	45.60	46.70	38.10	38.80	17.80
CHACALTAYA, BOLIVIA	0.55	0.43	2.31	--	2.45	8.72	3.09	5.94	3.10	3.88	--	1.60
ANTOFAGASTA, CHILE	6.83	0.51	7.57	7.42	4.35	--	4.64	--	--	5.34	4.74	29.50
PORTILLO, CHILE	4.71	8.06	--	--	--	--	--	--	--	--	--	--
SANTIAGO, CHILE	39.50	37.70	60.40	85.30	129.00	178.00	174.00	0.75	65.40	79.80	33.60	28.60
PUERTO MONTT, CHILE	0.48	--	--	--	--	--	1.66	0.95	0.53	0.05	--	--
PUNTA ARENAS, CHILE	5.84	4.17	4.41	5.91	6.79	8.61	11.30	5.74	3.76	4.03	--	2.00

-- NO DATA

Table 7 (Cont'd)

STABLE LEAD CONCENTRATIONS IN SURFACE AIR DURING 1970
($\mu g/m^3$) × 100

SITE	JAN.	FEB.	MAR.	APR.	MAY	JUNE	JULY	AUG.	SEP.	OCT.	NOV.	DEC.
NORD, GREENLAND	--	--	--	--	--	--	--	--	4.73	2.49	7.35	--
THULE, GREENLAND	2.86	0.92	0.70	0.10	0.40	0.59	0.73	0.68	1.03	0.92	1.20	--
MOOSONEE, ONTARIO	4.33	2.82	4.83	--	3.97	7.00	9.07	9.39	10.90	15.20	12.50	8.15
NEW YORK, NEW YORK	261.00	190.00	154.00	173.00	227.00	190.00	53.60	23.20	275.00	153.00	74.60	136.00
STERLING, VIRGINIA	47.60	62.20	52.40	47.40	37.90	31.10	61.00	27.70	33.40	44.10	83.20	39.40
MIAMI, FLORIDA	262.00	223.00	108.00	109.00	113.00	109.00	60.70	104.00	140.00	131.00	115.00	70.00
BIMINI, BAHAMAS	3.30	3.85	0.36	0.46	3.74	--	6.37	4.50	3.32	3.48	1.84	2.63
MAUNA LOA, HAWAII	--	1.22	1.27	0.67	1.57	0.52	4.97	0.49	0.36	0.27	0.41	1.12
SAN JUAN, PUERTO RICO	112.00	177.00	78.90	116.00	23.60	93.40	45.60	82.30	207.00	237.00	102.00	79.30
BALBOA, PANAMA	16.50	27.50	--	36.50	34.80	26.70	27.80	--	56.40	--	--	--
GUAYAQUIL, ECUADOR	69.90	67.30	53.10	--	54.70	47.20	22.30	12.40	35.20	21.60	28.40	17.40
LIMA, PERU	36.90	62.90	30.90	132.00	45.20	15.70	--	--	2.60	9.24	5.71	25.60
CHACALTAYA, BOLIVIA	1.17	1.57	1.59	0.63	4.58	--	--	--	2.64	2.79	1.66	1.67
ANTOFAGASTA, CHILE	6.10	2.14	3.62	--	--	--	--	6.18	5.10	1.75	5.80	8.23
SANTIAGO, CHILE	33.30	40.40	72.30	145.00	228.00	160.00	361.00	--	77.30	62.60	52.60	33.40
PUERTO MONTT, CHILE	--	--	--	--	--	3.46	0.30	2.73	2.68	4.55	4.68	2.78
PUNTA ARENAS, CHILE	4.08	4.23	5.03	2.67	2.15	86.60	14.80	9.75	11.70	10.70	7.50	9.28
GABRIEL GONZALEZ VIDELA	0.90	2.83	2.80	2.76	1.00	0.94	1.62	1.70	--	4.92	3.67	2.82
SOUTH POLE STATION	--	--	--	--	0.53	0.34	1.00	1.72	3.44	7.02	0.29	0.43

-- NO DATA

Table 7 (Cont'd)

STABLE LEAD CONCENTRATIONS IN SURFACE AIR DURING 1971
$(\mu g/m^3) \times 100$

SITE	JAN.	FEB.	MAR.	APR.	MAY	JUNE	JULY	AUG.	SEP.	OCT.	NOV.	DEC.
NORD, GREENLAND	2.59	--	--	--	--	--	--	--	--	--	--	--
THULE, GREENLAND	0.30	--	--	--	--	--	--	--	--	--	--	--
MOOSONEE, ONTARIO	5.84	--	--	--	--	--	--	--	--	--	--	--
NEW YORK, NEW YORK	133.00	--	--	--	--	--	--	--	--	--	--	--
SALT LAKE CITY, UTAH	331.00	--	--	--	--	--	--	--	--	--	--	--
STERLING, VIRGINIA	42.60	--	--	--	--	--	--	--	--	--	--	--
MIAMI, FLORIDA	177.00	--	--	--	--	--	--	--	--	--	--	--
BIMINI, BAHAMAS	12.10	--	--	--	--	--	--	--	--	--	--	--
MAUNA LOA, HAWAII	1.39	--	--	--	--	--	--	--	--	--	--	--
SAN JUAN, PUERTO RICO	76.60	--	--	--	--	--	--	--	--	--	--	--
GUAYAQUIL, ECUADOR	34.50	--	--	--	--	--	--	--	--	--	--	--
LIMA, PERU	37.70	--	--	--	--	--	--	--	--	--	--	--
CHACALTAYA, BOLIVIA	1.40	--	--	--	--	--	--	--	--	--	--	--
ANTOFAGASTA, CHILE	6.43	--	--	--	--	--	--	--	--	--	--	--
SANTIAGO, CHILE	65.10	--	--	--	--	--	--	--	--	--	--	--
PUERTO MONTT, CHILE	1.67	--	--	--	--	--	--	--	--	--	--	--
PUNTA ARENAS, CHILE	11.90	--	--	--	--	--	--	--	--	--	--	--
SOUTH POLE STATION	0.18	--	--	--	--	--	--	--	--	--	--	--

-- NO DATA

RESULTS

All of the available data on monthly lead concentrations from the surface air program are shown in Table 7. This table is abstracted from the quarterly report of the Health and Safety Laboratory[2]. The results are in units of micrograms of lead per standard cubic meter of air ($\mu g/m^3$) multiplied by 100 for convenience in tabulation.

The monthly trace metal deposition in total New York City fallout from December 1969 through May 1971 are presented in Table 8. The units are $\mu g/cm^2$. Note that we have only tabulated the results for vanadium, copper, zinc, lead and cadmium. Chromium, manganese and nickel have been omitted because the value in the blank approaches or exceeds the amounts measured in fallout, and the results are therefore probably meaningless. Undoubtedly, these high blank values simply reflect a degree of removal of metals from the stainless steel pot during evaporation and extraction of the sample at the end of the collection period.

TABLE 8. TRACE METALS IN TOTAL FALLOUT
IN NEW YORK CITY ($\mu g/cm^2$)

Date	V	Cu	Zn	Pb	Cd
1969					
Dec.	2.17	0.79	2.71	3.68	0.02
1970					
Jan.	1.76	0.66	2.30	1.91	0.02
Feb.	1.97	0.97	2.79	1.77	0.03
Mar.	2.25	0.82	2.88	3.61	0.03
Apr.	2.86	N.D.	N.D.	3.94	N.D.
May	1.18	N.D.	N.D.	2.69	N.D.
June	N.D.	0.82	4.60	3.81	0.03
July	N.D.	0.59	2.96	2.89	0.02
Aug.	0.94	1.08	2.67	3.29	0.02
Sept.	0.48	0.64	1.89	2.43	0.02
Oct.	1.05	0.89	1.73	1.91	0.02
Nov.	1.31	0.89	1.58	2.50	0.02
Dec.	2.29	0.87	3.31	3.15	0.02
1971					
Jan.	3.68	1.10	1.86	3.42	0.02
Feb.	3.68	0.50	2.21	1.63	0.02
Mar.	1.84	0.66	2.31	3.81	0.02
Apr.	1.37	0.87	2.96	3.42	0.03
May	1.25	0.72	2.44	2.89	0.03
June	0.69	0.93	2.49	2.78	0.02
Mean	1.81	0.81	2.57	2.92	0.02

Tables 9 and 10 list the values for trace metals in precipitation in New York for the period, Sept. 1970 through July 1971. Table 9 is in units of $\mu g/cm^2$ for easy comparison with the total fallout data of Table 8. Table 10 shows the same data in concentration units of micrograms per milliliter of precipitation ($\mu g/ml$) or parts per million (ppm). In these tables the values for cadmium have been omitted because the collection device apparently contaminated the samples with cadmium.

TABLE 9. TRACE METALS IN PRECIPITATION
IN NEW YORK CITY ($\mu g/cm^2$)

Date	V	Cr	Mn	Ni	Cu	Zn	Pb
1970							
Sept.	0.07	0.09	0.05	1.42	0.44	3.51	0.87
Oct.	0.15	0.10	0.11	N.D.	0.57	1.33	1.04
Nov.	0.55	0.05	0.08	0.24	0.38	0.98	0.89
Dec.	1.03	0.13	0.15	0.33	0.37	1.04	1.03
1971							
Jan.	1.09	0.21	0.11	0.37	0.33	2.87	0.95
Feb.	1.91	0.21	0.14	1.26	0.35	N.D.	1.23
Mar.	0.76	0.25	0.15	0.36	0.46	1.72	1.16
Apr.	0.46	0.11	N.D.	0.18	0.22	0.81	0.75
May	0.48	0.15	0.11	N.D.	0.35	2.73	1.09
June	0.21	0.14	0.11	0.52	0.41	1.68	0.75
July	0.41	0.19	0.16	0.74	0.44	2.43	1.03
Mean	0.65	0.15	0.12	0.60	0.39	1.91	0.98

N.D. = No data.

TABLE 10. CONCENTRATION OF TRACE METALS IN PRECIPITATION
IN NEW YORK CITY ($\mu g/ml$)

Date	V	Cr	Mn	Ni	Cu	Zn	Pb
1970							
Sept.	.013	.017	.010	.260	.080	.643	.160
Oct.	.024	.015	.017	N.D.	.088	.205	.160
Nov.	.051	.005	.008	.022	.035	.091	.083
Dec.	.183	.022	.027	.059	.066	.185	.183
1971							
Jan.	.236	.045	.024	.080	.071	.620	.207
Feb.	.161	.018	.012	.106	.030	.703	.104
Mar.	.096	.031	.019	.045	.058	.216	.146
Apr.	.127	.030	N.D.	.049	.060	.221	.206
May	N.D.	.019	.014	N.D.	.045	.346	.138
June	.093	.060	.049	.232	.183	.751	.337
July	.020	.010	.008	.037	.022	.121	.051
Mean	.100	.025	.019	.099·	.067	.373	.161

N.D. = No data.

DISCUSSION

Lead in Surface Air

The three and one half years of almost continuous data on lead concentrations in surface air has revealed a number of interesting trends and has suggested some conclusions which were not at all obvious at the outset.

Low Concentration Areas. As described earlier, the surface air program was originally set up for radioactive fallout studies, and the Pb analyses were added later. The sites are therefore not all very well suited for establishing the type of base line data desired; that is, air concentrations which represent a fairly large region over a long period of time. This is most evident at sites where very low concentrations were expected. One of the problems seems to be that a relatively small amount of vehicular traffic or gasoline operated equipment is sufficient to bias a sample. Hence, whereas we expected the results from Thule, Mauna Loa, Chacaltaya and the South Pole to indicate very constant low lead levels, they turn out quite variable (see Table 7). Thule, for example, varied from about 10^{-3} (almost our detection limit) to a high of 3×10^{-2} $\mu g/m^3$. Mauna Loa has shown even wider variation, as has the South Pole.

Similarly, the results from the Atlantic Ocean weather ships are probably elevated, to an unknown degree, by the effect of emissions from the ship itself, such as flaking of leaded paint, etc. Table 11 lists annual mean Pb concentrations for the surface air sites. Qualitatively, from these data, it appears that the Echo ship does in fact generally sample higher concentration air than the other three. Considering the geographic positions of these ships in the Atlantic Ocean (see Table 1) it may be that Bravo, Charlie and Delta could reasonably be considered to be in areas of very low lead concentration, and virtually all the lead measured is ship contamination. Echo, however, at 35° north latitude, adjacent to and generally downwind from the heavily populated east coast of the U. S. might be expected to exhibit real, tropospheric lead concentrations, in the measurable range.

The only set of consistent and probably realistic low results come from the sampler at Puerto Montt, Chile. This

site was located about ten miles out of the business part
of the city, in an area of very sparce automobile traffic.
Through 1969, mean annual Pb concentrations were between
4 and 7 x 10^{-3} $\mu g/m^3$. In mid 1970, the sampler was moved
to the city airport, a center of vehicular traffic, and the
1970 mean value increased by about a factor of four. These
average yearly concentrations are shown in Table 11.

TABLE 11. ANNUAL AVERAGES OF STABLE LEAD IN SURFACE AIR
$(\mu g/m^3)$
Stations arranged in order, North to South

Location	1967*	1968	1969	1970
Thule, Greenland	.009	.012	.014	.009
Atlantic Bravo (Ship)	.018	.041	.020	-
Atlantic Charlie (Ship)	.014	.020	.020	-
Moosoonee, Ontario	.085	.047	.12	.074
Atlantic Delta (Ship)	.017	.059	-	-
New York City	2.8	2.0	1.5	1.7
Sterling, Virginia	.72	.74	.56	.45
Atlantic Echo (Ship)	.062	.10	.11	-
Miami, Florida	2.4	1.6	1.8	1.4
Bimini, Bahamas	.064	.14	.056	.038
Mauna Loa, Hawaii	.018	.007	.007	.015
San Juan, Puerto Rico	1.1	1.0	1.4	1.4
Balboa, Panama	.23	.26	.26	.33
Guayaquil, Ecuador	.37	.35	.30	.43
Lima, Peru	.39	.47	.36	.43
Chacaltaya, Bolivia	.017	.022	.032	.020
Antofagasta, Chile	.068	.072	.079	.042
Santiago, Chile	.87	.93	.82	.99
Puerto Montt, Chile	.004	.005	.007	.027
Punta Arenas, Chile	.076	.060	.057	.15
South Pole	-	-	-	.02*

*Incomplete year of data.

High Concentration Areas. In Figure 2, the monthly
lead concentrations are plotted as a function of time for
the 4 highest level sites. New York City shows a non-
systematic variability over more than an order of magnitude,
while Miami, averaging a little less, does not vary as much.
Both of these cities seem to indicate slowly diminishing
Pb concentrations, with time. The cause of this decrease

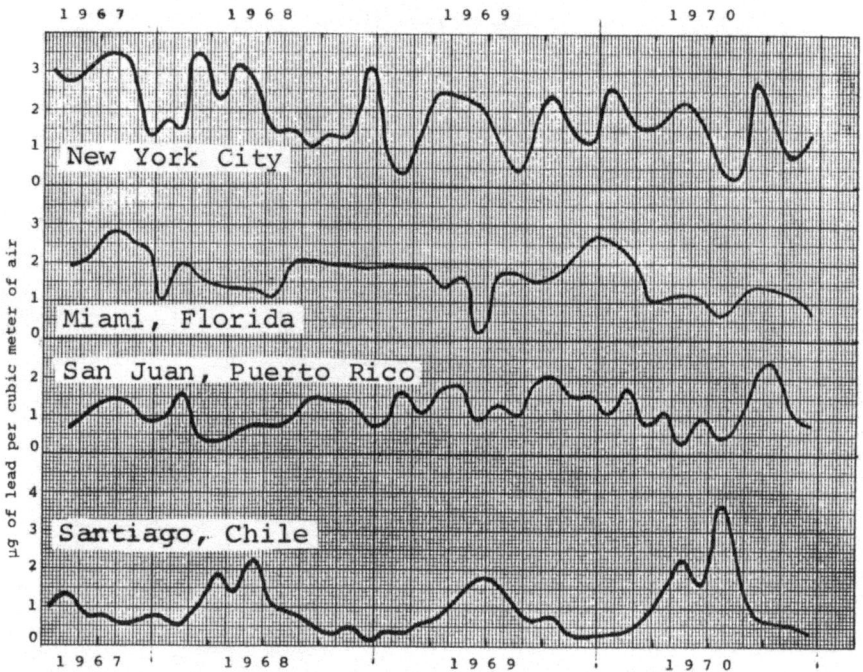

Figure 2.   Lead concentrations in surface air as a function of time.

is not understood at this time. San Juan, on the other
hand, averaging for the most part half or less than
New York in Pb concentration, is gradually increasing. The
data in Table 11, reinforce these conclusions. The average
in both Miami and New York approached a twofold decrease in
the period 1967 to 1970, while the value in San Juan has
increased in the same period and in 1970 was actually equal
to Miami.

        The only site exhibiting an obvious seasonal fluctua-
tion of the Pb in surface air, is Santiago, Chile. The
graph in Figure 2 clearly illustrates this. The high
occurs about mid-year, April through August, generally
during the Southern Hemisphere autumn and winter. The
peaking in part probably reflects a local meteorological
situation, the common occurrence of a low altitude inversion
over the city of Santago, during those months. These
inversions "trap" the air in and immediately over the city,

severly limiting mixing with air from higher altitudes;
hence the concentration of Pb emissions from vehicles
builds, until the inversion situation breaks, and fresher
air is brought in.

Trace Metals in New York City Fallout

On the basis of only 19 months of total fallout data
and less than a year of "wet" results we do not believe
that interpretation in depth can be very meaningful. Some
fairly obvious characteristics observed in the summaries of
Tables 8, 9 and 10 will be described in this section; and
where possible, our results will be compared to those of
other investigators.

Comparing trace metal concentrations in total fallout
and precipitation (Tables 8 and 9), the differences are
indeed seen to be significant. The ratio wet fallout/
total fallout for the four metals with comparable data,
indicate the following averages:

Vanadium	0.36
Copper	0.48
Zinc	0.74
Lead	0.34

These suggest that most of the zinc comes down in rain,
while less of the V, Cu and Pb are so deposited. Perhaps
this difference is related to the proximity of our sampler
to local sources of these metals. These ratios could also
be influenced by the chemistry of the metals or their
compounds in the atmosphere; whether or not a particular
form acts as an efficient condensation nucleus for precipitation would affect the ratio wet/total fallout.

A seasonal fluctuation of traces metal fallout in New
York City is only obvious for vanadium. In Table 8 it is
clearly evident that in the winter months the V fallout
reaches $2-3$ $\mu g/cm^2$ while in summer the level drops to less
than 1. The wet fallout (of Table 9) follows the same
pattern. This observation is in general agreement with an
earlier report[8] relating vanadium concentration in airborne
particulates to fuel oil consumption.

Precipitation samples from a nationwide sampling net-

work were analyzed for a number of metals covering the
period September 1966 to January 1967[9]. Table 12 summarizes these data and compares the averages and maxima with
the New York City averages from Tables 9 and 10.

TABLE 12. TRACE METALS IN PRECIPITATION –
NEW YORK CITY AND U.S.A.

	Deposition ($\mu g/cm^2$)			Concentration ($\mu g/l$)		
	Maximum USA[a]	Average USA[a]	NYC[b]	Maximum USA[a]	Average USA[a]	NYC[c]
Pb	1.38	0.23	0.98	150	34	161
Cu	0.54	0.15	0.39	68	21	67
Zn	4.96	0.62	1.91	200	107	373
Mn	0.31	0.08	0.12	50	12	19
Ni	0.14	0.03	0.60	12	4	99

[a]From reference 9.
[b]From Table 9.
[c]From Table 10.

As expected, both the deposition and concentration of these
trace metals in New York City are generally much higher than
the U. S. A. averages. Only manganese appears to be an
average contaminant in New York. The results for nickel
certainly seem to be excessive. At this point, no explanation for this discrepancy is offered; we can see no obvious
flaw in either our sampling or analytical procedure to
presume serious error in these data.

Collection of total fallout on the roof of the R.C.A.
building, 72 stories (or 260 m) above street level, were
initiated in March of 1971. To date only four months of
data have been completed, and with the problems always
encountered in setting up a new station the results should
not be taken without reservation. Nevertheless, the ratio
HASL Roof/RCA Roof (38 m vs. 260 m) averaged over the 4
months are as follows:

Vanadium   1.12
Copper   0.50
Zinc   0.90
Lead   1.39

At face value, these ratios would indicate that the important source of lead in total fallout is at low elevation, such as auto exhaust, and a measurable depletion of this contaminant in aerosols can be observed with height above street level. For copper, the opposite is suggested and we suspect contamination as the cause.

Engelmann[10] discussed the use of washout ratios as a means of predicting the scavenging of particles by precipitation. He defined the mass washout ratio as:

$$\frac{\text{material/gm water}}{\text{material/gm air}} = \left(\frac{k}{x}\right)_m$$

"material" denotes amount and may be in units of pCi, gms, etc.

The ratios reported by Engelmann for a large variety of substances varied quite widely, but for particulate debris the range was from about 100 to 2000.

Strontium-90 washout ratios computed from data in the HASL fallout and surface air programs are quite constant, averaging about $600\pm20\%$. The Pb washout ratios averaged about the same, but were much more variable with a standard deviation of about $\pm150\%$. This month to month variability of the Pb washout ratio must be the result of short term meteorological conditions, directly affecting the concentration in precipitation and surface air; and the agreement of the average with Sr^{90} is thought to be coincidental.

CONCLUSION

On balance, considering the time and effort expended on this program to date, we conclude that these studies should be continued and improved to provide some of the information necessary for scientific evaluation of trace metal problems in the environment.

The base line type data advanced in our surface air program already indicate trends of increasing or decreasing concentrations of Pb at several of the sites. At others, particularly at low lead levels, the interference caused by very local sources and samplers, bias the analytical results beyond usefulness as valid regional samples.

The multiple trace metals studies in New York City fallout are not sufficiently developed to provide more than very preliminary and general conclusions. Our data indicate that approximately 1/3 to 3/4 of the metals studied are deposited by precipitation, the balance is dry fallout. Not surprisingly, the trace metal fallout in New York, except for manganese, is substantially higher than United States averages; nickel far exceeds any maximum value reported earlier, and must therefore be examined closely.

We continue to see the need for improved samplers to cut down on contamination and reduce the detection limits. Improvements in the separation of wet and dry fallout collections, and plans for operating such devices unattended at remote sites are already under way.

REFERENCES

1. Sr^{90} and Sr^{89} in Monthly Deposition at World Land Sites, Hardy, E. P., Jr., Editor, USAEC Report HASL-245, Appendix pg. A-1 (October 1971).

2. Volchok, H. L. and Kleinman, M. T., Radionuclides and Lead in Surface Air, USAEC Report HASL-245, Appendix pg. C-1 (October 1971).

3. Krey, P. W. and Kleinman, M. T., Project Airstream, USAEC Report HASL-245, pg. II-8 (October 1971).

4. Krey, P. W. and Kleinman, M. T., High Altitude Balloon Sampling Program, USAEC Report, HASL-245, pg. II-37 (October 1971).

5. Bennett, B. G., Sr^{90} in the Diet, USAEC Report HASL-242, pg. I-108 (April 1971).

6. Freudenthal, P. C., Sr^{90} Concentrations in Surface Air: North America vs. Atlantic Ocean from 1966 to 1969, J. Geophys. Res., _75_, pg. 4089 (1970).

7. Kleinman, M. T. and Volchok, H. L., The Quality of Analyses in the HASL Surface Air Sampling Program During 1969, USAEC Report HASL-239 (January 1971).

8. Kneip, T. J., Eisenbud, M., Strehlow, C., and

Freudenthal, P. C., Airborne Particulates in New York
City, J. of Air Pollution Control Assoc., 20, pg. 144
(1970).

9. Lazrus, A. L., Lorange, E., and Lodge, J. P., Jr.,
    Lead and Other Metal Ions in United States Precipita-
    tion, Environ. Sci. Tech., 4, pg. 55 (1970).

10. Engelmann, R. J., Scavenging Prediction Using Ratios
    of Concentrations in Air and Precipitation, Proceedings
    of the 1970 Precipitation Scavenging Conference,
    Richland, Washington. CONF-700601, Natl. Tech. Info.
    Service, U. S. Dept. of Commerce, Springfield,
    Virginia 22151 (1970).

TRACE METAL CONCENTRATION FACTORS IN AQUATIC ECOSYSTEMS

Theo. J. Kneip and Gerald J. Lauer

New York University Medical Center
550 First Avenue
New York, New York 10016

ABSTRACT

The concept of concentration factors in aquatic systems is of ultimate importance in assessing the impact of any contaminant on such an ecosystem. While the general concept of bioconcentration is widely used, it is readily apparent that the qualifying assumptions and detailed theory underlying the concept often are not considered, leading to widespread misuse and misunderstanding of the phenomenon.

The basis is presented for the understanding of bioconcentration and concentration factors, and the (major) qualifying assumptions are discussed.

A series of examples are presented which demonstrate the variations in the bioconcentration of trace metals. These examples include elements known or thought to be essential as well as those known to be adventitious. The differences in the behavior of the metals demonstrate the problems inherent in general statements regarding concentration factors, and the reasons that many misuses have occurred.

INTRODUCTION

Metal uptake and concentration by food chains leading to man is a topic of renewed interest and concern because

of several occasions of unexpected human intoxications.
Mercury, cadmium, and lead are all implicated in such
problems (1, 2, 3). This reawakened interest has led to
a number of generalized statements regarding the concentration of metals by biota, and the possibility of a magni-
fying effect by repeated transfer upward through aquatic
food chains. That of Wolman (4) is typical, "Because
radioactivity is carried upward in the food chain and
because of the long half-life of many radionuclides,
continuing vigilance appears warranted." The conclusion is
most certainly correct, but the statement of food chain
activity is misleading. Again Wolman states "In many ways,
the chlorinated hydrocarbons pose much the same problem as
radioactive materials do; they are persistent and concentrations increase from water, to sediment, and thence to
biota with progressively higher concentrations higher in
the food chain." There appear to be numerous exceptions
to this statement as it applies to chlorinated hydrocarbons
(5), and the case for such food chain magnification of
radionuclides and metals in aquatic ecosystems appears weak
or non-existent.

Perhaps the first need is to clarify certain definitions. As defined by Polikarpov (6); a concentration
factor is the ratio of the concentration of a material in
a biological species to the concentration of the material
in water or the preceding link in the food chain. Emphasis
has been placed by Polikarpov on the need for steady state
or equilibrium conditions in determing such factors. Total
substance concentrations are normally used without regard
to variations in chemical form.

The concentration factor cannot be interchangeably used
with such terms as biomagnification and bioaccumulation.
These three terms can be differentiated in the following way:

Bioconcentration (concentration factor) refers
to the ability of an organism or a population of
many organisms of the same trophic level to concentrate a substance from an aquatic system.

Bioaccumulation refers to the ability of an
organism to not only concentrate, but to continue
to concentrate essentially throughout its active
metabolic lifetime, such that the concentration
factor, if calculated, would be continuously

increasing during its lifetime.

Biomagnification is the term which should be used when a substance is found to exist at successively higher concentrations with increasing trophic levels in ecosystem food chains.

Thus, the term biomagnification should be used only when a definite food web relationship has been demonstrated.

GENERAL RELATIONSHIPS

There are many major variables which may affect concentration factors. Discussion of a few of them serves to illustrate the problems in the use of such factors. It must be recognized that for the concentration factor to have meaning at any time other than the instant at which the concentrations are initially determined, organisms must develop a steady state or equilibrium concentration in their tissues relative to environmental exposure levels. Therefore, in every dynamic system the concentration factors must be determined throughout the range of variation of all of the significant variables in order to obtain a full understanding of the long term meaning of the concentration factors. Significant factors in dynamic systems include biological cycles, the response time of organisms vs. the time relationship of variations in input, seasonal effects, and possibly even diurnal variations in the system.

A major problem in use of organism concentration factors based on water concentrations is the almost universal existence of multiple routes of exposure. Any given organism may obtain a substance of interest from the food it ingests, indirectly from sediment ingested with or in place of food, through ingestion of water, and by transfer across external membranes. As the concentration factor takes into account the ratio of the concentration in biota to that in water only, it is obvious that other exposures may confuse the relationship between the water and the organism.

In some cases it has been shown that a concentration factor demonstrates some proportionality to the concentration of the substance in the ambient water, thus no constant concentration factor exists (7). This problem can also

occur because of a tendency for the distribution among the
organs within an organism to vary depending on inherent
physiological pathways, exposure concentration and/or the
stage in the biological cycle in which the organism exists
at the time of exposure.

BIOLOGICAL FACTORS

The concentration of a substance in biota may be
affected by many of the complexities of aquatic ecosystems.
The selectivity of organisms for the food they ingest and
the differing concentrations in the various food organisms
serve as an example. The differences in quantities of food
or water ingested in relation to the organisms age, the
season, and the biological cycle are further complicating
factors. If the availability of a substance to the predator
is affected by organ distribution or biochemical form in
food organisms, ingestion of such organisms will affect the
efficiency of transport across digestive system membranes.
Furthermore, utilization of food organisms (and presumably
other ingested substances) varies widely with need and
availability.

Retention and excretion are also major factors.
Generally, the retention of the material is limited by
physiological equilibria considerations. If an ion exchange
process is involved in the retention of the material, the
number of sites available to any substance will be limited;
and therefore, an absolute maximum of total material
retained will be observed. This can be confirmed as the
actual mechanism by determining whether a dead cell or dead
organism has the same capacity for uptake as a living
organism or cell.

In the case of metabolic limitations, either a toxic
factor or a physiological equilibrium factor may be involved
in establishing an equilibrium or steady state concentration
in the living organism. The factors will obviously vary
with organism, biological cycle, season and the element or
substance in question.

KINETIC FACTORS

Two major factors are involved in the kinetics with
which equilibria or steady states are established between

the medium and the given organism. The concentration
already present in an organism will obviously be involved
in establishing the gradiant across membranes, and therefore,
will be a major term in any equation which describes the
rate of uptake of a substance. An organism which has had no
exposure to a material will obviously take it up, if it can,
at a more rapid rate than one which has been heavily exposed.

The rate of excretion from an organism containing a
substance will also affect the rate at which the equilibrium
or steady state concentration is achieved. Thus, if no
excretion occurs, or if it is very slow compared to the
uptake rate, bioaccumulation will be observed whereby the
concentration in the organism continues to increase throughout its lifetime.

Organism size, i.e., surface to volume ratio, may also
be of considerable importance in the determination of the
rate of uptake. The smaller the organism, in general, the
higher the surface to volume ratio, and therefore, the more
rapid will be transport across the surface as compared to
the total mass of organism involved. A second factor is
that large organisms may generally have lower rates of
metabolism, or age related metabolic rates which can effect
changes in the kinetic factors with age and size.

It is obvious that the time of duration of the exposure
is of significance. For materials which are ingested, the
digestion of the food stuff and/or conversion of the substance involved to a utilizable form may well be involved
in determining the absorption of any given material. All
of these then come into play in determining the transfer
coefficient, which is the fraction of the material crossing
the membrane as compared to the ambient available mass
involved in the exposure.

BIOCHEMICAL EFFECTS

Trace elements may be divided into three major classes.
These are essential elements, toxic elements, and accumulated elements. While this is an oversimplification, it
does provide a first order differentiation of the elements
according to their biochemical effects.

Essential elements, if present in sufficient amounts,

will have no effect other than those beneficial uses which
define the essential nature. If insufficient amounts are
present, deficiency diseases may occur as biological effects,
and essential elements may exhibit toxic effects at excessive
exposure concentrations.

Any toxic element may have two major classes of effect.
Those are the lethal effects and chronic sublethal effects.
Most lethal or acute effects have been defined for toxic
metals in aquatic systems for many years with a resulting
decrease in interest in such studies until fairly recently.
Recognition that chronic sublethal effects might be occurring in individual organisms and on whole populations has
led to an awakening of interest in studies of such problems.

Accumulated elements may of course eventually reach a
toxic level even though the exposure may be constant and
extremely low. Even though no toxic effect occurs in the
organism accumulating the element, an effect may appear at
a higher trophic level because of ingestion of the accumulating organism as a food stuff.

Interactions of various elements may have marked effects
on the apparent toxicity. Many metals exhibit lower toxicity
in hard water as compared to soft, and this is generally
interpreted as caused by a competition with calcium. The
higher calcium concentrations in hard water blocks the
action by filling transfer sites or by other undefined
mechanisms. The well known reduction of cadmium toxicity
with elevated zinc concentrations is a second example. Thus
a detailed knowledge of biochemical mechanisms is required
to predict the many possible biochemical effects and interactions.

It will only be through a thorough understanding of
the actual ecosystem functions that we can define the degree
to which elements may be put in circulation by man's
activities without long term, potentially irreversible
changes in significant ecosystems.

FOOD WEB DISTRIBUTIONS

A series of examples have been chosen to illustrate the
behavior of different chemicals or elements as they have
been observed in existing aquatic ecosystems. The examples

have been chosen by two major criteria. Sampling was done
simultaneously at all levels of the food web, and sufficient
information was provided that a reasonable probability exists
that steady-state or equilibrium conditions existed.

The materials which have undergone most intensive
study are the organo-chlorine pesticides. Biomagnification of these materials during passage through aquatic
food webs appears to have been decisively demonstrated
by a number of careful studies. However, recent controlled
laboratory studies on the kinetics of DDT uptake by gold-
fish indicate that some observations of apparent biomagnification might actually be caused by the kinetics of
uptake relative to the length of time organisms are
exposed to environmental contaminants (8 and 9). Ecosystem
data on pesticides has been quoted in such a way as to
give the impression that almost all materials and
elements are biomagnified in every aquatic system. Actually
considerable doubt exists as to the occurrence of biomagnification throughout the entire aquatic food web
even for chlorinated hydrocarbons (5). It will become
obvious through the examples quoted here that this is not
a general behavior for most elements.

ORGANIC COMPOUNDS

Chlorinated Hydrocarbons - A variety of data has been
published relating to the concentration of pesticides in
aquatic ecosystems. Several sets of this data have been
selected for presentation in Figure 1 to demonstrate that
these oil soluble, water insoluble, chlorinated hydrocarbon compounds show apparent biomagnification in food
webs, with calculated concentration factors of 100,000 or
more in the organisms at the top of the food web. It is
readily seen in the figure that the concentrations in
sediments, plants, plankton and bivalves are similar,
while fish, birds, and predatory birds show increasing
concentrations. Thus biomagnification is important only
at the upper trophic levels. The concentration factors
calculated for fish, birds and predatory birds are
obviously artificial in that the data demonstrates that
the food stuff has become an important route of exposure,
and the ratio of concentration in the biota to the
concentration in the water is therefore a purely artificial number. Among these organisms only fish can be

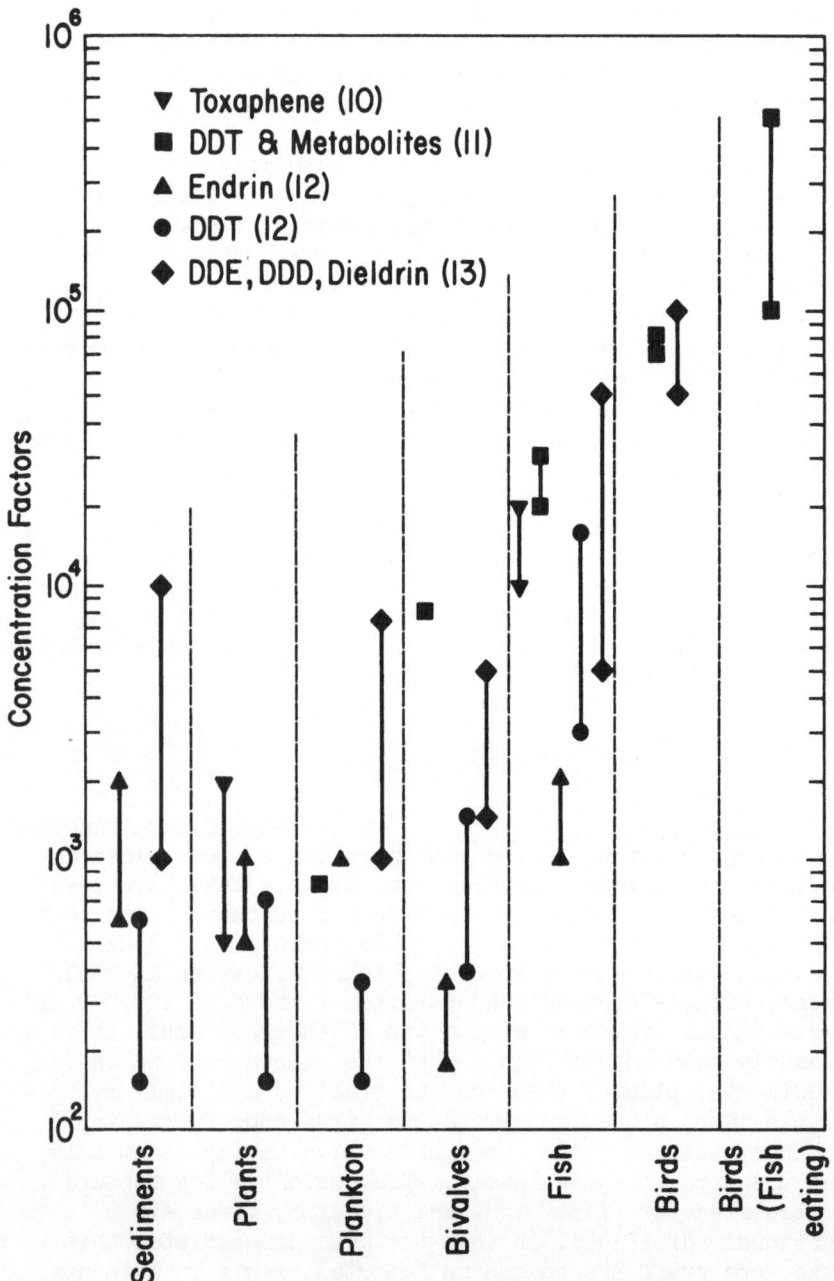

Figure 1. Pesticide Concentration Factors

expected to have water as a major source of exposure.

Methyl Mercury - Organo-mercury compounds have been studied in relation to the recent mercury poisoning problems in human populations. While a large amount of information has been gathered on the concentration of mercury in various organisms, very few studies have simultaneously sampled and analyzed for mercury in several trophic levels in the food web, and essentially none have been performed or reported where the concentration in water was determined at the same time that the concentrations in the biota were observed. The data from two studies are shown in Figure 2. Hannerz' (14) data has been converted to concentration factors based on the mercury concentrations which were produced by addition of mercury to water in these laboratory studies. The data of Johnels (15) could not be converted to concentration factors as the concentration in water was never determined in that study. It is perhaps fortunate that these two sets of data can be plotted on the same axis since the field observations of Johnel's, in µg/g tissue, fall in the same range of absolute numbers as the concentration factors calculated from Hannerz' data.

As seen in the figure, some tendency toward bioaccumulation may exist in passing from worms through insects in the food web. However, the data taken in the field do not indicate a clear and certainly not a significant biomagnification from worms through insects to fish. While the overall data available (but not reported here) indicates that the top predatory fish have the highest concentrations, there is still insufficient information to determine the extent to which biomagnification occurs, as distinguished from bioaccumulation by larger (and therefore older) individuals in species of the top trophic level. There is no question that bioconcentration of the organo-mercury compounds does occur, but it is obvious that biomagnification, if present, is many orders of magnitude less significant than that observed for the oil-soluble organo-chlorine pesticides. This may relate to the major storage of the mercury as a complex in muscles with binding to sulfhydryl sites in proteins.

ESSENTIAL ELEMENTS

Manganese - An example of concentration factors for an

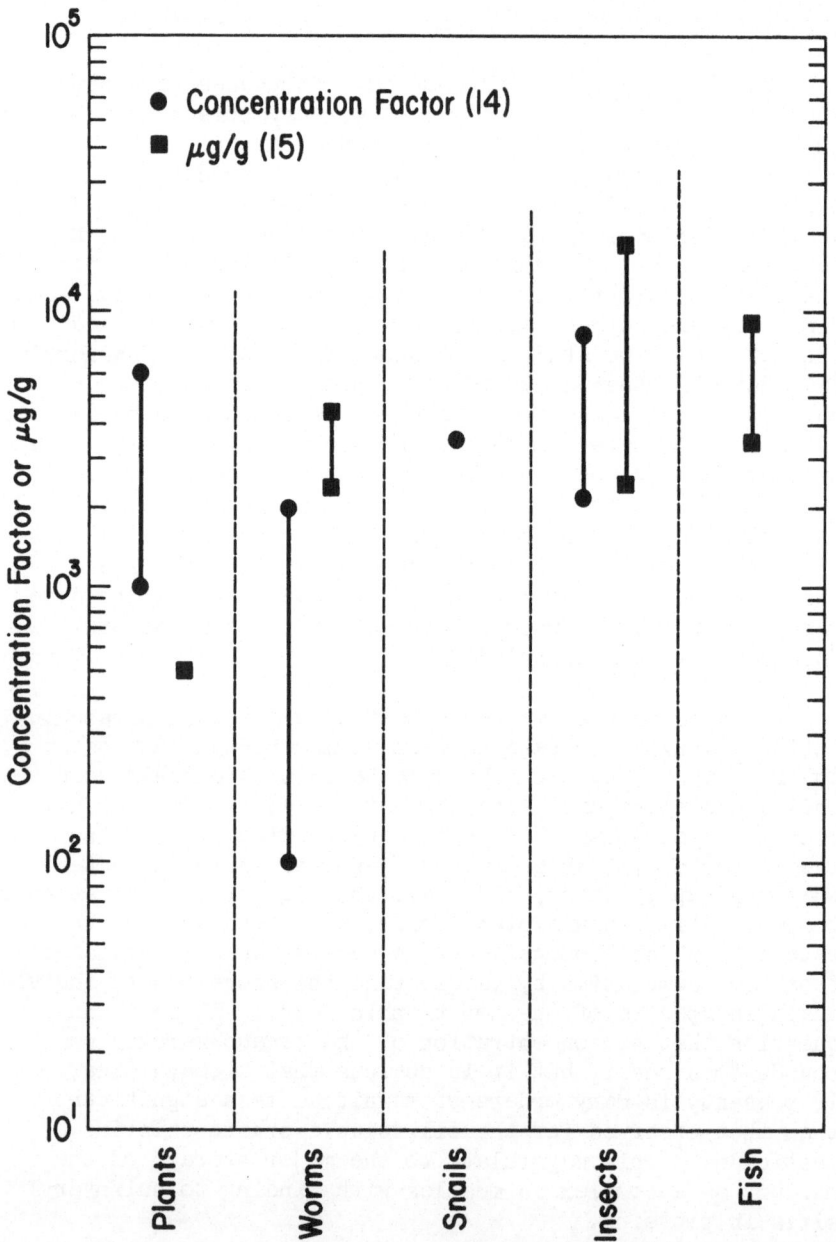

Figure 2.   Methyl Mercury Distribution

essential metal in an estuarine system is shown in Figure 3
(16). It is obvious that the lowest trophic level in this
system has achieved the greatest concentration factor while
the most predatory organism noted, the striped bass, has
the lowest concentration factor. It is generally observed
that the higher organisms are capable of regulating the
element manganese; and therefore, do not have higher
concentration factors. It is also evident that biomagnification is not occurring in this system at the time of the
year that the sampling was performed.

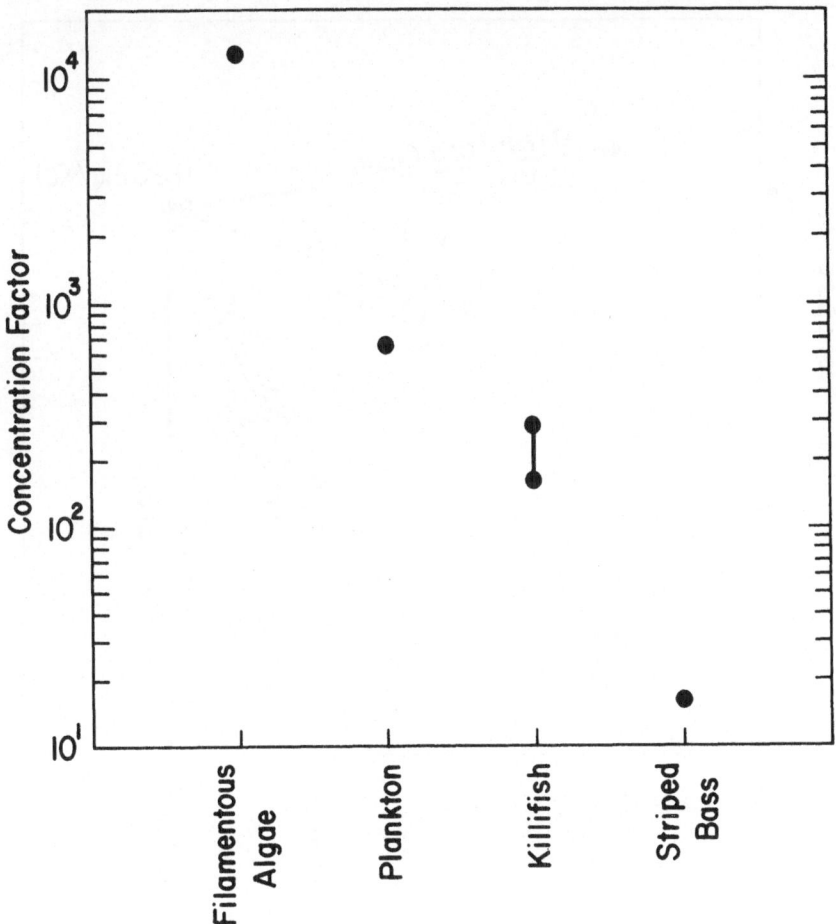

Figure 3.   Concentration Factors for Manganese in Hudson
River Aquatic Food Chain--June, 1970

Additional data for manganese in the system are presented in Figure 4 indicating the concentration factors and
food web relationships between bivalves and several sources
of exposure. Had the bivalve been placed on the previous
figure, it would have had a concentration factor approximately as high as the filamentous algae and would have been
placed in the food web between the plankton and the Killifish indicating an up and down relationship in the overall
system. However, it is obvious that the bivalve cannot have
a meaningful simple concentration factor as it has at least

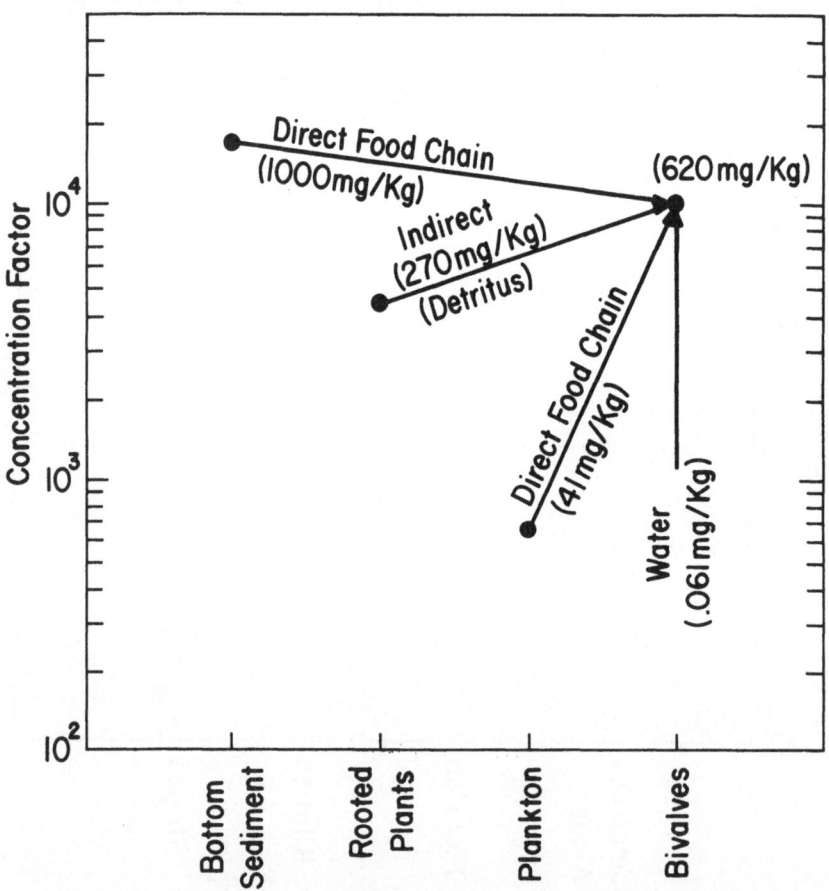

Figure 4. Manganese: Concentration Factors and Food Web
            Relationships

four modes of input, and there is no effective way of
determining the actual concentration present in the organism.
Thus, for the bottom sediment source alone, the concentration factor calculated would be less than 1 while that for
water is 10,800. The values calculated for the indirect
or direct food chain materials, such as detritus or plankton,
would range from slightly greater than 10 to slightly greater
than 1. The complexity is obvious in this particular case.

The concentration of manganese in the water has
been noted to vary rapidly in the Hudson River estuarine
system because of exchange between water and bottom
sediments. Such dynamic factors must be carefully accounted
for in obtaining and interpreting the data as the rate of
establishment of steady state concentrations varies with
trophic level.

Copper has been selected as a second example of an
essential element. Concentration factors have been calculated from the data of Wilber (17) for several species
using the normal sea water concentration as a base (Figure
5). It is readily seen that an apparent concentration
occurs between algae and invertebrates, but does not
continue to the higher level of food chain represented by
fish. The need to define the direct food input is obvious
as the fish sampled and examined may not have been foraging
on the invertebrates analyzed in this particular study. The
data does not indicate that copper is biomagnified, while
it is obvious that it is bioconcentrated.

The importance of copper and the critical nature of
the element to oysters can be noted from the data for sea
water vs. the critical limits for successful oyster cul-
turing shown in Figure 6. The range for successful reproduction is from 7 to approximately 170 times the concentration in sea water, while only 1.7 times more than the
concentration which maximizes reproduction results in
toxicity and failure to reproduce.

Adventitious Elements

Cesium - The element cesium has been studied at some
length to define the behavior of the radionuclides produced
by atomic testing or nuclear reactor power stations. In
Figure 7 a set of data for the Hudson River organisms is
presented (19). No biomagnification can be observed in

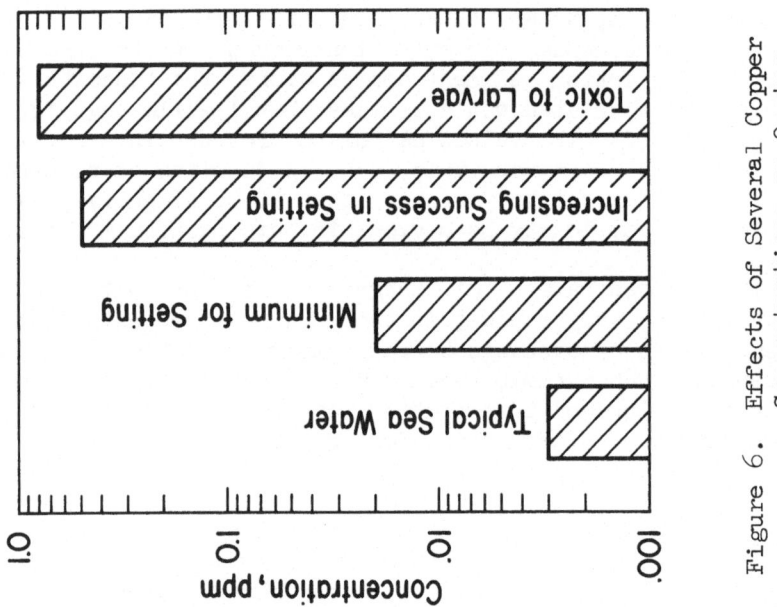

Figure 6.   Effects of Several Copper
            Concentrations on Oyster
            Reproduction

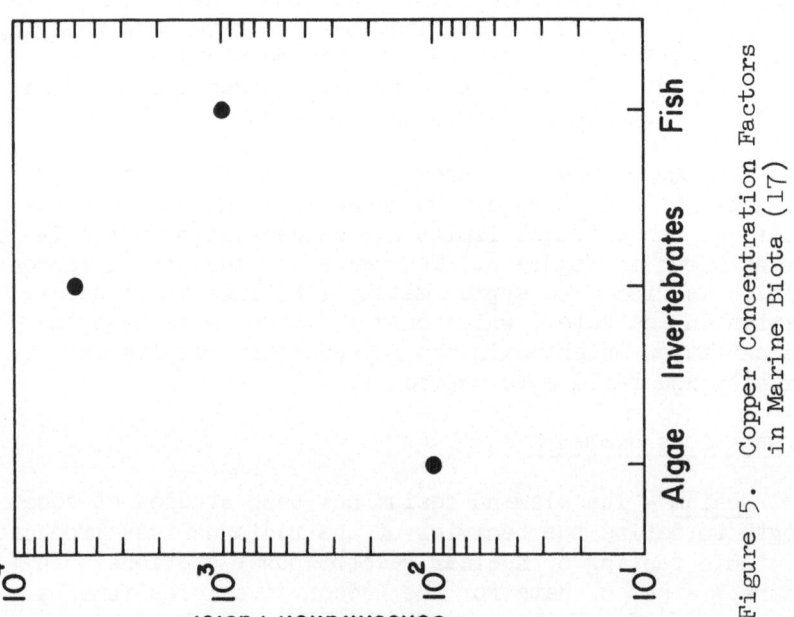

Figure 5.   Copper Concentration Factors
            in Marine Biota (17)

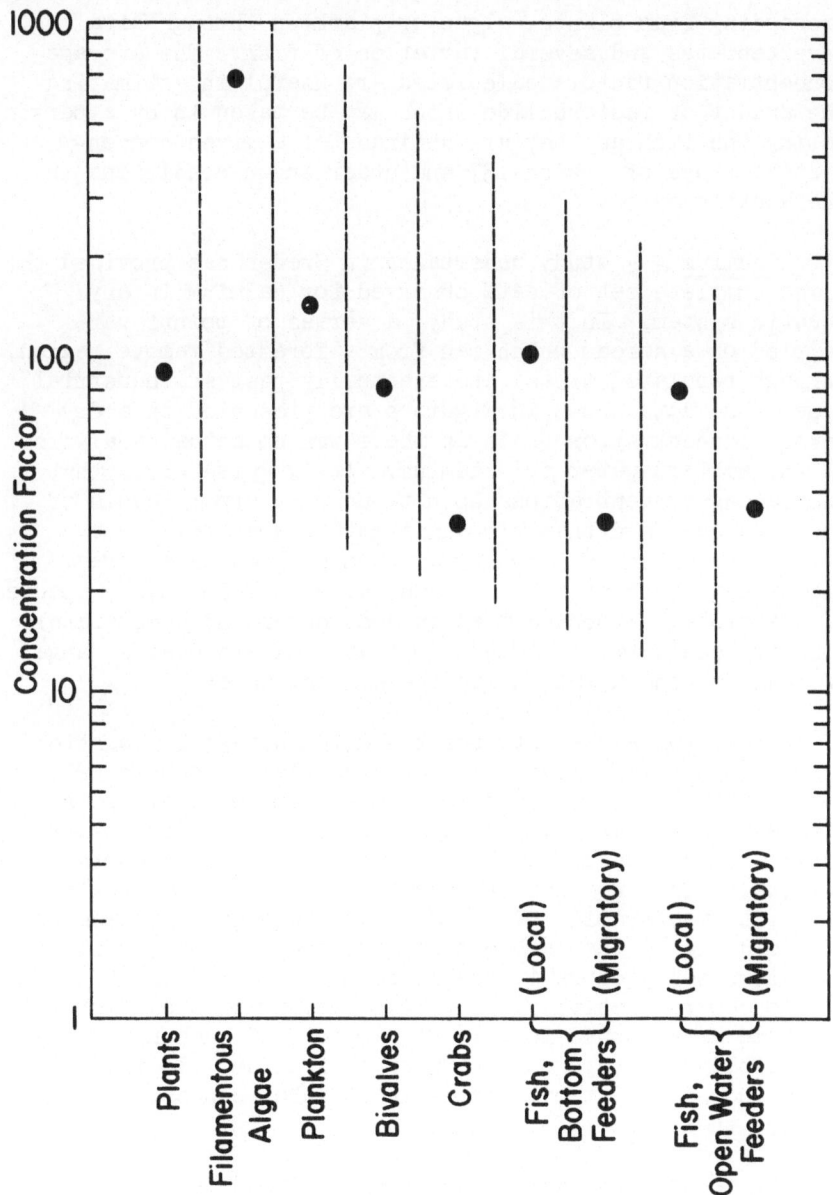

Figure 7.   Concentration Factors for CS-137 in Hudson
           River Aquatic Food Chain--June, 1970

proceeding from plants, algae or plankton through bivalves, invertebrates and several varieties of fish. The average concentration factors calculated are useful in estimating the amount of radionuclide which may be taken in by a person eating the fish as they are defined for a given concentration range of cesium-137 and under known conditions in the aquatic system.

Cadmium - A study undertaken in Sweden has provided the first complete set of data observed for cadmium in any aquatic system. In this study, a series of points were sampled on a stream which ran from a forested remote region, through populated areas, and eventually past an industrial zone. The data shown in Figure 8 are presented on a dry weight concentration basis as there was no actual analysis of the ambient water for cadmium. It is quite clear that increasing concentrations do not occur at upper levels of the food web, and that biomagnification does not occur. At site 6, which was downstream from an industrial area, the highest concentration of cadmium occurs for every species of the biota. Bioconcentration does occur and a moderately constant concentration factor probably exists over a range of cadmium concentrations in the ambient water.

Plutonium - The data for plutonium uptake by aquatic organisms has been reviewed most recently by Noshkin (21). Concentration factors reported range from 20 to 21,000 for marine invertebrates and algae. Several relationships are apparent from the data. Plutonium shows higher factors for shells than for soft tissues in all cases. Soft tissues of filter feeders in 1970 studies off Cape Cod were 200 to 400 fold higher than ambient water concentrations, while concentration factors in shells from the same organisms were 300 to 600 respectively. Starfish feeding on shellfish in the same area showed factors of 760 to 4100, demonstrating a possible one step food chain magnification. Zooplankton, algae and seaweed tended to exhibit higher concentration factors than the higher organisms.

Concentration factors for fish reviewed in the same study were in the ranges of 1-13 for muscle, 14-175 for liver and 21-570 for bone. Gut membranes were examined in three fish and found to show concentration factors of 36,730 and 1060. This later data is stated to indicate a relation between gut absorption and physico-chemical form of the plutonium.

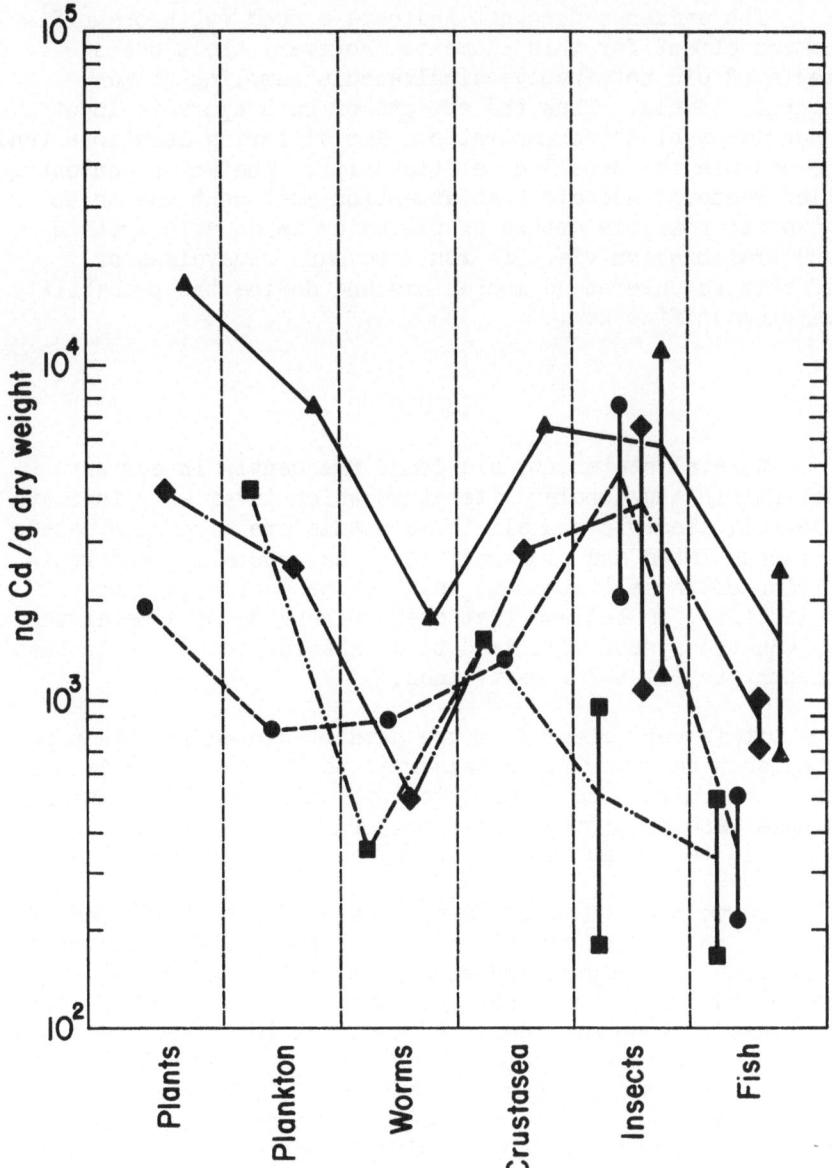

River Mörrumsån, Sweden (19)
Sampling Locations. 3(■); 4(●); 5(◆); 6(▲)

Figure 8. Cadmium Concentrations in a Fresh Water Food Web

The evidence does not indicate a food web biomagnification effect for this element. However, the studies
reviewed did not involve simultaneous sampling of many
trophic levels. Thus the effects of both sporadic inputs,
changing ambient concentrations and differing locations tend
to obscure the detailed relationships. The known concentration factors indicate that attention must continue to be
given to possible uptake of plutonium in aquatic systems
and transmission via this route to man, regardless of
efforts to understand mechanisms and define the probability
of biomagnification.

DISCUSSION

General statements are found frequently in current
scientific and popular literature which invariably indicate
that all elements and all toxic metals are bioconcentrated,
bioaccumulated and biomagnified. The generality and repi-
tition of these statements has led the public, and many
scientists, to believe that every discharge of a metal to
an aquatic system will lead to a behavior identical to that
thought to exist for pesticides.

It is very clear from the data presented in this paper
that such behavior for metals must be the exception rather
than the rule. The information presented here includes no
single case of biomagnification. This was as surprising
a result to the authors as it may be to the readers.

Surveys of biota for metals as currently performed are
usually focused on organisms used as foodstuffs, with other
organisms, sediments, and water are either ignored or
neglected (often reluctantly because of cost or methodology
problems). Adequate means for eventual long term solutions
of these problems will not be achieved on the basis of
general statements which are incorrect. It is essential
that careful efforts be made to examine existing data,
generate thorough environmental studies, and evaluate the
real probability of effects due to bioconcentration, bioaccumulation, or biomagnification. A considerable effort
remains to be performed before long range predictions of
ecological effects of metals may be made.

REFERENCES

1. Department of Health, Education and Welfare, Special
    Report to the Secretary's Pesticide Advisory Committee,
    "Hazards of Mercury," Environmental Research, 4(1),
    (March, 1971).
2. Friberg, L., M. Piscator and G. Nordberg. "Cadmium in
    the Environment," 153, C. R. C. Press, Ohio, (1971).
3. National Academy of Sciences. Lead--Airborne Lead in
    Perspective, National Academy of Sciences, Washington,
    D. C., (1972).
4. Wolman, M. Gordon. "The Nation's Rivers," Science,
    174:905-918, (November, 1971).
5. Edwards, Clive A. Persistent Pesticides in the Environ-
    ment, C. R. C. Press, Cleveland, Ohio, (1970).
6. Polikarpov, G. G. Radioecology of Aquatic Organisms,
    Reinhold Book Division, New York, (1966).
7. Butler, Philip A. "Pesticide Residues in Estuarine
    Mollusks," presented at the National Symposium on
    Estuarine Pollution, Stanford University, August, 1967
    (sponsored by the American Society of Civil Engineers).
8. Grzenda, Alfred R., Doris Fort Paris and William J.
    Taylor. "The Uptake, Metabolism and Elimination of
    Chlorinated Residues by Goldfish (Carassium auratus)
    Fed A ^{14}C-DDT Contaminated Diet," Transactions of the
    American Fisheries Society, 99(2), (April, 1970).
9. Youngs, W. D., W. H. Gutenmann and D. J. Lisk.
    "Residues of DDT in Lake Trout as a Function of Age,"
    Environmental Science and Technology, 6(1):451-452,
    (May, 1972).
10. Terriere, L. C., U. Kiigemagi, A. R. Gerlach and R.
    L. Borovica. "The Persistence of Toxaphene in Lake
    Water and its Uptake by Aquatic Plants and Animals,"
    J. Agr. Food Chem., 14(1):66-69, (1966).
11. Woodwell, G. M., C. F. Wurster, Jr. and P. A. Isaacson.
    "DDT Residues in an East Coast Estuary: A Case of
    Biological Concentration of a Persistent Insecticide,"
    Science, 156:821-824 (1967).
12. Godsil, P. J. and W. C. Johnson. "Residues in Fish,
    Wildlife, and Estuaries," Pesticides Monitoring Journal,
    1(4):21-26, (1968).
13. Kneip, T. J., unpublished data (1969).
14. Hannerz, L. "Experimental Investigations on the Accumu-
    lation of Mercury Compounds in Water Organisms," Rep.
    Inst. Freshwater Res., 48:120, (1968).
15. Johnels, A. G., M. Olsson and T. Westermark. "Esox

Lucius and Some Other Organisms as Indicators of
Mercury Contamination in Swedish Lakes and Rivers,"
Bull. Office Int. Epiz., 69(1439), (1968).

16. Lentsch, J. W., T. J. Kneip, M. E. Wrenn, G. P. Howells
and M. Eisenbud. "Stable Manganese and Manganese-54
Distributions in the Physical and Biological Components
of the Hudson River Estuary," Proc. Third National
Symposium on Radioecology, Oak Ridge (In press, 1972).

17. Wilber, C. G. The Biological Aspects of Water Pollution,
Charles C. Thomas, Springfield, Illinois, 216, (1969).

18. Prytherch, H. F. "The Role of Copper in the Setting,
Metamorphosis and Distribution of the American Oyster,"
Ostrea virginica. Ecological Monographs, 4(1):47-107,
(1934).

19. Wrenn, M. E., J. W. Lentsch, M. Eisenbud, G. J. Lauer
and G. P. Howells. "Radiocesium Distribution in Water,
Sediment, and Biota in the Hudson River Estuary from
1964 through 1970," Proc. Third National Symposium on
Radioecology, Oak Ridge (In press, 1972).

20. Ljunggren, K., B. Sjostrand, A. G. Johnels, M. Olsson,
G. Otterlind and T. Westermark. "Activation Analysis
of Mercury and Other Environmental Pollutants in Water
and Aquatic Ecosystems," Nuclear Techniques in Environmental Pollution, Proceedings of a Symposium, Salzburg,
26-30 October, 1970, International Atomic Energy Agency,
Vienna, (1971).

21. Noshkin, V. E. "Ecological Aspects of Plutonium Dissemination in Aquatic Environments; What Has Pu Data
to Tell us About other Transuranics," Presented at the
11th Hanford Biology Symposium, The Biological Implications of the Transuranium Elements," Richland, Washington, 26-29 September, 1971, (In press).

Acknowledgment

Support for aquatic research studies has been provided
by the U. S. Public Health Service Bureau of Radiological
Health, the New York State Department of Health, and the
Consolidated Edison Company of New York. The project is
part of a Center program supported by the National Institute
of Environmental Health Sciences, Grant No. ES 00260.

GWYNETH HOWELLS

NATURAL ENVIRONMENT RESEARCH COUNCIL
LONDON   UK

INTRODUCTION

There is a long-standing interest in the composition of
man, for a variety of reasons. One, of course, is a natural
curiosity and an intuition that the nature of man might be
revealed by his composition. This concept has been ration-
alised to conceive that it might be possible to identify
cultural, ethnic or environmental differences between com-
munities or population groups. Another reason for an
interest in tissue composition or "body burden" has been the
association of some specific diseases or physiological
manifestations in man with perturbations of tissue concentrations from the "normal".

A further reason for our interest is the present concern for the effects of trace metals in the environment of
man: every few months it seems that it is suggested that a
new element is a hazardous pollutant and that levels in
human tissue or in food/air/water are reaching dangerous
levels. Some of these fears derive from, and are extra-
polations of, concern for occupational risks. For instance,
lead, cadmium, arsenic and beryllium are known from
industrial experience to be hazardous. But evidence has
been accumulating that lead, and in some circumstances
mercury and cadmium are _environmental_ (in contrast to
occupational) risks. This implies that they are hazardous
to the general population at the sort of concentration
levels which occur in the general environment.

The controversies which arise, involving questions of
field observation and the identification of specific
effects, are bedevilled by lack of data. We simply do not
know enough about the levels of these substances in the
environment, about the forms in which they are present,
about the processes, and the dynamic aspects of these processes, which modify or transfer them through the environment, or about their biological effects and the nature of
the dose-effect relationship. It is typical that
"manifestations" of the effects of these pollutants in man
often depend on subjective symptoms, such as central
nervous system effects which include tiredness, depression
and irritability, and also show great individual variation.

One of the ways in which such environmental hazards
may be brought into perspective is to review the natural
(or "normal") state of affairs which can establish a base-
line, with its natural variability, against which unnatural
perturbations can be measured. On the one hand it is
necessary to document the normal levels, forms, sources
and sinks of "pollutants" in the natural environment, and
to study the pathways of these materials in organisms,
including man, and their effects.

BODY COMPOSITION OF MAN

From this argument it can be accepted that the composition of man is an important and worthwhile objective for
study. Although important, however, it is an aim difficult
to achieve. Only a very few whole bodies (apart from
still-born babies) have been analysed (1). These few are
sufficient to indicate that there is considerable variation
due to age, sex, occupation, environment and the immediate
past history of health, as well as more fundamental differences due to genetic or ethnic differences. It is clear
that the few data available cannot provide for a standard
or "reference" base-line.

TABLE 1

COMPOSITION OF WHOLE BODIES (70 Kg)

(Widdowson and Dickerson 1964)

Oxygen	44,800 g
Hydrogen	5,600 g
Nitrogen	2,380 g
Calcium	1,568 g
Phosphorus	840 g
Potassium	188 g
Sodium	129 g
Chlorine	124 g
Magnesium	33 g
Iron	5.2 g
Zinc	1.9 g
Copper	120 mg
Boron	28 mg
Cobalt	1.5 mg

Some techniques are available for studying composition
"in vivo". These are by tracing exchangeable water (2),
exchangeable sodium (3), or similar substances (4) where the
body can be considered as a "single component" with a
relatively simple dynamic system. However, even with such
mobile materials of general distribution as water and the
sodium ion, it has become clear from these studies that
"exchangeable sodium" for instance is not necessarily
"total sodium" in the context of the experiment.

A more certain technique is to estimate the proportion
of natural or induced, long lived radioactive nuclides in
the body. This can be done for the natural isotope of
potassium K-40 (5), or for a few of the long lived fallout
nuclides which might be considered to have reached an
equilibrium. However, the technique of whole-body counting
of radionuclides at these low environmental levels is only
applicable to very few isotopes.

There has been the recent development of whole-body
neutron activation (6) which induces short lived radioactive nuclides which can subsequently be measured. This
technique is limited to observations of those elements which
yield detectable and conveniently counted nuclides, but also
by the strict limitation on radiation dose that must be
imposed on neutron activation of volunteer subjects.

The procedure which generally has to be adopted might
be termed the "jig-saw" technique which assembles a
"reference man" from analysis of tissues from a wide range
of samples to yield values for the concentration of substances. The "reference man" is then constructed from
tissue quantities and concentrations. A further sophisti-
cation is to define the turnovers of these substances from
information on input, output and retention, or from
experimental studies.

STANDARD MAN

The International Commission for Radiation Protection
(ICRP) identified a need for a "standard man" some 15 or 20
years ago at the Harriman House conference. A "standard" of
value limited by the available data, was set up and published
in the 1959 report "Permissible Dose for Internal Radiation"
(Publication 2). Recognition of inadequacies of that data,
and the possibilities offered by new techniques, led to the

setting up in 1963, of a "Standard Man Task Group" * who
were to define the anatomy, composition and physiology of a
standard man, with indications of the variations due to age,
body size and sex. A manual of values for "Reference Man"
is being prepared for publication shortly.

The philosophy behind the concept of "standard man"
was that information about the distribution and concentration of stable elements, and of the normal input, output
and retention, could be used to interpret and predict the
behaviour of the radioactive nuclides which might be introduced to the body. While this argument is still valid for
long-lived isotopes in environmental equilibrium or for
short-lived nuclides carried by "bulk" natural elements, it
may now be possible to provide the information by more
direct methods. None-the-less, "standard" or "reference"
man will continue to be important as a base-line reference
point for investigations of radionuclides in the body, and
also of other contaminating substances. In addition, the
anatomical and physiological data are essential for the
sophisticated radiation dose calculations which are now
called for in the evaluation of environmental or occupational radiation hazards.

PROBLEMS

The formulation of a "reference man" by the jig-saw
method has been fraught with a number of problems, not all
of them resolvable with present technology since some are
conceptual and not to be verified by observation or experiment. Analysis of tissues, for the most part, has been
done by the technique of spark source mass-spectrometry and
hence reveals elemental composition, whereas it becomes
increasingly evident that the forms of substances are of
great importance. Some elements, also, are not amenable to
this technique which requires preliminary ashing of samples,
since they, or their compounds, or their association in
tissues, are volatile to a greater or lesser extent.

*   Dr W S Snyder (Chairman), Oak Ridge National Laboratory
    Miss Mary Jane Cook,   Oak Ridge National Laboratory
    Dr Lucien Karhausen, CEA, Paris
    Dr Gwyneth Parry Howells, Medical Research Council, London
    Dr Isobel Tipton, University of Tennessee

In this case, alternative techniques are used, but may have
differing analytical precision. In some instances, e.g.
for micro-trace elements, existing techniques provide only
information of the "less than level of detection" variety.
For some elements, a proportion of tissue samples may
indicate measureable levels, the remainder, a less than
detection level; this presents a problem of data analysis
where a "normal standard and range of variation" is sought.
In addition, some element concentrations in tissues do not
exhibit a normal range of variation, nor a consistent
pattern which might be amenable to mathematical handling.
In many instances, the pattern of distribution of data is
unknown because of the limited number of samples analysed.

The origin and handling of tissues, especially for
trace element analysis, poses more familiar problems. The
"standard man" tissues have been derived from accident
cases, not hospital patients, but this has meant that the
past history of the body is rarely known. One can only
hope and assume that the victims of accidents represent a
cross-section of a typical population, not selected in any
way which would influence tissue composition specifically,
even though all cases may not have been in optimal health.
The data are valid for a population sample. The most
rigorous precautions have been made in the collection,
separation and laboratory handling of tissues to avoid any
spurious contamination.

The establishment of a "reference" anatomy has posed
different, but no less intractable problems. Much of the
literature comes from the last century and is quite
acceptable, but some anatomical concepts were not formulated with present day precision. For instance, it is
true but may be surprising to acknowledge, that lung
anatomy was very imperfectly known until very recent years.
Since the different parts of the lung (tracheal, bronchial
and alveolar) respond physiologically in different ways to
inhaled particles and vapours, it is necessary to define
these parts anatomically in terms of tissue mass, surface
area and volume. "Lung tissue" as presented to the
analyst represents the dissected lungs from the bifurcation
of the trachea and includes the bronchial tree, pulmonary
lymph nodes and capillary blood as well as the lung paren-
chyma; venous and arterial blood will have been lost at
dissection. However, in compiling the "reference man",
values for the lost blood would have to be included.

Other tissues, which may have a very specific composition in relation to trace substances, may be difficult to isolate or define. Examples are body fat (present as depot fat, total fat, adipose tissue, essential intercellular fat etc.) and bone with its associated but separate tissues of periosteum, tendon, cartilages, blood, red and fatty marrow etc. When these difficulties in obtaining data are compounded with the need to define the differences due to sex, age and growth, occupation or environment, the task can be seen as difficult indeed.

The physiological reference values have also suffered from a lack of relevant information and from an unevenness in the data available. Values were required of the overall intake of food, water and air, and for the appropriate levels of metabolism. While air (and thus O intakes and CO_2 exhaled) is closely related to metabolism and activity and carefully documented, food and water intakes are much less reliable.

Food intake can be determined in a number of ways - from national consumption and population estimates, from market basket surveys or questionnaire type dietary surveys, or by controlled balance studies of a few individuals. The first has the disadvantage that the age structure of the population is rarely recorded, that techniques of sampling are notoriously difficult and may be variable in different studies (e.g. family food consumption or family food purchase), that meals or snacks taken outside the household are not recorded (but may or may not be allowed for) and that single elements are rarely studied but are derived by calculation from standard tables of food composition not necessarily applicable to the situation in question. On the other hand, these estimates have the advantage that large populations can be studied, and the consumption "per caput" can be checked against national supplies of food components. The balance study has the obvious disadvantages that few individuals can be studied and that free chosen diets often show considerable idiosyncrasy of choice. The advantages are that age and sex are usually recorded, the whole food intake can be followed, and generally specific elements are measured in the food by analysis of aliquots or duplicate meals. The two methods however often provide widely different estimates of intake (7).

Water intake, and the associated intake of dissolved
elements, is woefully inadequate in the literature, except
for extreme environmental conditions which are hardly
relevant to the defined reference situation. In addition,
water composition may vary widely with location, and in turn
may influence or be influenced by the availability of other
materials.

Urinary excretion of elements is measured by analysis
of samples, usually of a relatively small number of individuals, although there have been studies on a population
basis for a few elements (8). The extrapolation of values
obtained for a urine sample to a daily excretion will lead
to some error, and few sources relate the sample to daily
creatinine excretion or to time of sample collection; for a
few substances information is available on significant
circadian fluctuations. At least for those elements homeo-
statically controlled by the body, urinary concentrations
usually fall within a limited range, especially for normal
or typical diets, and this generally means more accurate
analysis of the sample. Consequently, where incompati-
bility of intake and excretion data have been found for the
reference man model, more reliance has been placed on
urinary analyses (where these are adequate). However,
interlaboratory studies of trace element (and even major
electrolyte) analysis of identical urine samples indicate
large variations, even when a common technique is
employed (9).

Estimates of excretion of elements in feces have been
derived from direct chemical analyses, not by difference
of intake and urinary excretion, but the data are limited.
The great variability of concentrations, reflecting both
variable intake and variable alimentary absorption, together
with great variation in the frequency of defecation, means
that the values in the literature generally show a very wide
range. Most analyses reported refer to stool concentrations
rather than to the daily losses needed for the reference
model. Where possible, values should be related to a
specific level of intake.

To these sources of variation in all samples must be
added the variability inherent in collating material provided by different investigators using different analytical
procedures on different subgroups of the trans Atlantic
population selected. The possibility of variation due to

differences of ethnic or regional groups should be considered whenever the model values are used for a particular
population. In selecting model values from available data,
an assumption has been made that average daily input and
average daily output are in balance, whereas day to day
fluctuations are known to occur.

STANDARD OR REFERENCE VALUES

Data derived by these admittedly inadequate means
have been accumulated into a "reference man". Somewhat
surprisingly the resulting "man" does not fall too far
short of expectations or of what is known by other techniques. The compilation can be broken down by the distinction between bulk, major and minor elements. An analysis
of the input or output as a fraction or per cent of body
burden is of interest but rarely indicates the metabolic
handling of elements by the body. A selection of data is
included in Table 2.

TABLE 2: NOTES

The table includes body burden data only for those
elements whose concentration is known in at least 50% of
the weight of total body tissue, including the skeleton.

F & F = food and fluids /day

U = Urine loss /day

F = Fecal loss / day

Ex = Exhaled /day

Air intakes were calculated as the upper limit due to
intake of urban air on the basis of a total daily air
intake of 2.3×10^4 litres.

TABLE 2

VALUES FOR BODY BURDEN, INTAKE AND OUTPUT OF ELEMENTS BY
"REFERENCE MAN". (70 Kg, daily basis)

Ele-ment	Units	Body Burden	Intake F & F	Intake Air	Output U	Output F
O	g	43 000	2 600	920	1 300	100
C	g	16 000	300	-	5	7 270 (Ex)
H	g	7 000	350	-	160	13 177 (others)
N	g	1 800	16	-	15	1.5
Ca	g	1 000	1.1	-	0.18	0.74
P	g	780	1.4	-	0.9	0.5
S	g	140	0.85	0.00054	0.8	0.14
K	g	140	3.3	-	2.8	0.36
Na	g	100	4.4	-	3.3	0.1
Cl	g	95	5.2	-	4.4	0.05
Mg	g	19	0.34	-	0.13	0.21
Si	g	18	0.0035	(0.00015?)	0.01	0.01
Fe	g	4.2	0.016	0.00003	0.00025	0.015
F	mg	2 600	1.8	-	1.0	0.15
Zn	mg	2 300	13	< 0.1	0.5	11
Rb	mg	320	2.2	-	1.9	0.3
Sr	mg	320	1.9	-	0.34	1.5
Br	mg	200	7.5	-	7.0	0.07
Pb	mg	120	0.44	0.01	0.045	0.3

Element	Units	Body Burden	Intake		Output	
			F & F	Air	U	F
Cu	mg	72	3.5	0.02	0.05	3.4
Al	mg	61	45	0.10	0.10	43
Cd	mg	50	0.15	< 0.001	0.1	0.05
B	mg	< 48	1.3	-	1.0	0.27
Ba	mg	22	0.75	0.00009	0.05	0.69
Sn	mg	< 17	4.0	0.00034	0.02	3.5
Mn	mg	12	3.7	0.002	0.03	3.6
I	mg	11	0.20	0.0005 - 0.035	0.17	0.02
Ni	mg	< 10	0.40	0.0006	0.011	0.37
Au	mg	< 10	-	-	-	-
Mo	μg	< 9 300	300	< 0.1	150	120
Cr	μg	< 1 800	150	0.1	70	80
Cs	μg	1 500	10	0.025	9.0	< 1.0
Co	μg	1 500	300	< 0.1	200	90
U	μg	90	1.9	0.007	0.05 - 0.50	1.4 - 1.8
Be	μg	36	12	< 0.01	1.0	10
Ra	μg	0.000031	2.3×10^{-6}	-	0.08×10^{-6}	2.2×10^{-6}

The present need for assessment of environmental hazards has emphasized the need for wide-ranging information on the "normal" levels of potentially toxic materials, including heavy metals, and their pathways through the environment and in man. Tables 3 - 7 summarise information on body burden, organs of high concentration, portals of entry, and routes of excretion which can be retrieved from the ICRP Reference Man Reports. It is natural that a review of this nature has revealed the inadequacy of much of the available information; hopefully the identification of the gaps will help to focus effort to remedy them.

The values for input of elements (as illustrated in Tables 3 - 7) are useful in evaluating the relative importance of foods, water or air in the total intake. For example, of the reference total daily intake of mercury in the "transatlantic" diet (10 μg), only 2 μg in UK could be derived from contaminated fish or fish products (10). Fluorine, often the subject of controversy when it is proposed as an additive to drinking water, is derived more from tea and beer than from fluoridated water. Excretion of potentially toxic materials derived from the diet or environmental exposure can also be an important source of pollution. Most mercury in the diet is unabsorbed and excreted via feces. Not surprisingly sewage sludge is high in heavy metal content, even in non-industrial areas, and sites where sludge in dumped routinely have built up high deposits.

The Reference Man information, besides serving its primary purpose for radioactive dose calculations, will help to provide an assessment needed for environmental hazards.

CONCLUSIONS

Needs for the future are, first a greater quantity of relevant and compatible information about body composition and body or organ contents of trace metals. Secondly, more needs to be known about the relative importance of different routes of entry, but especially of that part derived from inhalation. A most significant field of endeavour is to determine the availability of materials - in relation to the different portals of entry, in terms of the physical and chemical form, in relation to differences in age, sex or other physiological conditions, and in relation to differences in diet.

Within the body, we know very little about the dynamics
of transfer of materials from one organ or tissue to
another. Even whole-body turnover is known for only a hand-
ful of elements. We need also to study the effects of
disease and the range of variation for the normal situation.

Advances in the field will come rapidly with the
exploitation of the new techniques of analysis; even so the
task is large. Both an overall survey approach to the problem, and the more specifically directed study of the body
handling of an individual element or its compounds have
their place, together with the study of disease associated
abnormalities in trace-element metabolism, or the study of
extreme (occupational or accidental) exposures. Only by
bringing together all the information from diverse sources
will we be able to judge the true hazards of trace materials
in the environment.

TABLE 3

SUMMARY DATA FOR MERCURY

Body Burden: Not known for whole body, 0.013 g in soft
                                                tissue

Intake: Food and fluids 10 μg/day (No value for
                                  water)
             Inhaled 0 - 50 μg/day (0.01 - 760 μg/m^3 Air)

I/BB%:   0.08%

Output: Urine 0 - 35 μg/day (80% have no detectable
                                                Hg)

U/F: very small, 1.3/25 29 days after Hg^{203}

Turnover: 28 - 70 days (oral, Hg salts)
             70 - 200 days (fish in diet, methyl mercury)

Absorption: In gut, not known

Sweat: Present, concentration not known

Hair: Present, not closely related to dose or
                                              exposure

TABLE 4

SUMMARY DATA FOR LEAD

Body Burden: 0.12 grams, most in skeleton

Intake: Food and Fluids 0.44 mg/day
               (0.024 - 0.01 in water/day)
               Inhaled 0.01 - 0.1/day in towns
                 ($3\mu g/m^3$ air) 1 μg/cigarette

I/BB%:   Less than 0.4%

Output: Urine 0.045 mg/day
               Feces 0.3 mg/day, mostly unabsorbed food lead,
                              some endogenous excretion in bile

U/F: 0.15

Absorption: 5 - 14% from food in gut
                  10% from water or beer
               10 - 20% from air

Sweat: 0.065 mg/day (3 men, balance study)
               0.002 - 0.046 mg/50 cm^2 skin
               Concentration reflects exposure

Hair: 1 - 19 mg/100 g
               Reflects dose or exposure

TABLE 5

SUMMARY DATA FOR CADMIUM

Body Burden: 0.050 grams, most in kidney

Intake: Food + fluids 150 μg/day, high in shellfish/
                      grains. 0.3% of BB
              Water intake 12 μg/day, Water 0.006 μg/l
              Inhaled   < 1 μg/day, Air   0.002 - 0.1 μg/m^3

Output: Urine 100 μg/day (range 2 - 110 from balance
                                                study)
              Feces 50 μg/day
              U/F   2

Absorption   0.5 - 8% (as $CdCl_2$ in mice)
  in gut: < 12% (as a salt in rats)

Hair: 0.23 μg/100 g

              Accumulates in tissue with age

TABLE 6

SUMMARY DATA FOR NICKEL

Body Burden: Less than 0.01 grams

Intake: Food + fluids 400 μg/day (in most foods)
                      4% of BB
              Fluids   14μg/day - USA rivers 1 - 30 μg/l
                            UK drinking 150 μg/l
              Inhaled   0.6 μg/day, Air 0.03 - 4 μg/m^3

Output: Urine 11 μg/day (5% of Intake)
              Feces 370 μg/day
              U/F   0.003

Sweat: Present, said to be 83 μg/day (3 men balance
                  study) reference value taken as 20 μg/day

Hair: Present, 0.25 μg/100 g

TABLE 7

SUMMARY DATA FOR IODINE

Body Burden: 0.011 grams, most in thyroid

Intake: Food + fluids 200 μg/day ca. 2% of B.B.
                Water - variable geographically,
                        0.7 - 50 μg/l in UK
                Inhaled 0.5 - 35 μg/day Air 0.003 μg/m^3
                                              (sea level)
                                              Near sea 1-2 μg/m^3

Output: Urine 170 μg/day, approximates Intake
                Feces   20 μg/day, represents endogenous I
                U/F   8.5 or more

Absorption: "Complete"

Sweat: present 0.5 - 1.2 μg/100 ml, 6 μg/day

Saliva: 10 μg/100 ml

REFERENCES

1. E M WIDDOWSON and J W T DICKERSON (1964) "Chemical
    Composition of the Body", Chapter 17, pp 1 - 247, in
    Mineral Metabolism, Academic Press, New York and
    London, Volume II, Part A, edited by C L Comar and
    F Bronner.

2. G von HEVESY and E HOFER (1934) "Die Verweilzeit des
    Wassers im mensclichen Korper, untersucht mit Hilfe
    von "Schwerem" Wasser als Indicator" Klin. Wschr. 13:
    1524 - 1526

3. G B FORBES (1962) "Sodium", Chapter 25, pp 1 - 172, in
    Mineral Metabolism, Academic Press, New York and London,
    Volume II, Part B, edited by C L Comar and F Bronner.

4. F D MOORE, K H OLESEN, J D M^{C}MURREY, H V PARKER,
    M R BALL and C M BOYDEN (1963) The Body Cell Mass and
    its supporting Environment,W B Saunders and Co.,
    Philadelphia and London.

5. K T WOODWARD, T T TRUJILLO, R L SCRUCH and E C ANDERSON
    (1956) "Correlation of total body potassium with body
    water". Nature 178: 97 - 98

6. J ANDERSON, S B OSBORN, R W S TOMLINSON, D NEWTON,
    J RUNDO, L SALMON and J W SMITH (1964) "Neutron
    analysis in man in vivo". Lancet (ii) pp 1201 - 1205

7. M E GROOVER Jr., L BOONE, P C HOUK and S WOLF (1967)
    Problems in quantitation of dietary surveys.
    J.A.M.A. 201: 8 - 10

8. L J GOLDWATER (1967) "Normal" concentrations of metals
    in urine and blood. WHO Chronicle, pp 191 - 192
    L J GOLDWATER, M B JACOBS and A C LADD (1962) Absorption
    and excretion of mercury in man. I. Relationship of
    mercury in blood and urine. Arch. Environmental Health
    5: 537 - 541

9. M B JACOBS, A C LADD and L J GOLDWATER (1964) Absorption
    and excretion of mercury in man. VI Significance of
    mercury in urine. Arch. Environmental Health 9:
    454 - 463

10.   H M STATIONERY OFFICE (LONDON) (1971) "Survey of
      Mercury in Food" Report of a Working Party on the
      Monitoring of Foodstuffs for Mercury and other
      Heavy Metals; First report.

RELATION OF TRACE METALS TO HUMAN HEALTH EFFECTS.

Henry A. Schroeder, M.D. and Dan K. Darrow, M.S.
Department of Physiology,
Dartmouth Medical School
Trace Element Laboratory
9 Belmont Ave.,Brattleboro,Vt. and
Brattleboro Memorial Hospital

For the past 150 years, civilized man has been exposed
more and more widely to metallic contaminants in his environment, resulting from the products of industry. Canning of
foods, for example, was introduced in the Napoleonic Wars,
but did not become wide-spread until the Civil War. Smelt-
ing of ores and refining of metals has been going on a long
time, introducing metals into air and water, but human
exposures were usually local; during the past 50 years they
have become fairly general. Exposures to lead have occurred
in circumscribed areas of the world for 3000 years or more,
and were high among the Roman upper classes; the use of lead
pipes in soft water areas has lead to sporadic episodes of
lead poisoning in persons drinking these waters, but not
until 1924, when alkyl lead was put into gasoline as an
anti-knock agent were whole civilized populations exposed
to lead at an annually increasing rate. Cadmium was an
industrial curiosity in 1900, but today its use is sharply
increasing in an exponential curve, with resultant contamination of air, water and food. Mercury has been widely
used for amalgamation of gold from crushed ore, but discovery
of its catalytic and fungicidal properties has resulted in
considerable local contamination from seeds and dumping
effluents into stagnant lakes. Antimony was used as a
cosmetic by Cretan women, but now it is everywhere in glazes,
enamels and type metal. Almost every civilized person is
exposed to silver, gold, vanadium, chromium, titanium, nickel,
germanium, arsenic, selenium, tellurium, niobium, zirconium,
barium, in amounts exceeding those to which his forebears

were exposed, and those to which wild mammals are exposed
today. The earth is rapidly becoming a place where few
human beings can be found who are living at background
environmental levels.

As a result, the human body burden of many elements is
considerably increased over that of primitive man. The
question naturally arises: have any of these elemental substances recondite toxicity expressed as metabolic break-
down resulting in disease or as slow metabolic deterioration
resulting in decreased longevity?

Suspicion falls on any element which accumulates in
human tissues with age. Of the trace elements essential to
life, health and optimal function--vanadium, chromium, manganese, iron, cobalt, copper, possibly nickel, zinc,
selenium, strontium, molybdenum, fluorine, iodine--none
accumulate under present exposures, except in unusual and
individual situations where homeostatic mechanisms for
repulsion of excesses and conservation of deficiencies are
genetically disturbed, or when exposures are extremely large.
Copper--and iron--storage diseases occur in man as genetic
traits.

The non-essential trace elements found in appreciable
concentrations in the environment can be divided into those
with high natural levels in the earth's crust--silicon,
aluminum, titanium, barium, zirconium, niobium, lithium,
lanthanum, gallium--and those with low natural levels but
with high potential for industrial contamination. As mammals
developed in the presence of high environmental concentrations of the first group, one can assume--and show experi-
mentally--that they have low orders of toxicity.

Elements found in low concentrations on the earth's
crust but mined and used industrially in large quantities are
tin, germanium, beryllium, arsenic, molybdenum, mercury, antimony, bismuth, cadmium, silver, selenium, gold, tellurium.
As these elements also occur in low concentrations in sea
water, life developed and mammals evolved without the need
of elaborate mechanisms for handling them. Therefore, when
today environmental exposures increase many fold, some of
them accumulate in human tissues and exert recondite toxicity,
for the balance between absorption and excretion is disturbed.

All substances are toxic in large enough amounts, that

is, when homeostatic mechanisms for excretion are overcome.
In fact, water is one of the most toxic substances known,
when one compares the normal intake to the toxic intake.
A man can drink 3 liters a day, but if he drinks 9 or 10
liters, his kidneys may fail. A person can eat 10 to 15
grams of sodium chloride a day without adverse effects, but
if he eats 40 g he becomes edematous. The difference between
the normal intake of a trace element by food, water and air
and the toxic intake depends upon the efficiency of homeostatic mechanisms for excretion, but except for a few
instances, is several orders of magnitude higher.

A rough idea of present day exposures to the products
of industrial metals can be found by tabulating approximate
annual U.S. industrial consumptions in terms of the elements
necessary for life, those with known toxicities, those
slightly toxic and those inert.(Table 1). Of the 16 essential ones, present exposures offer no hazard whatsoever.
High exposures in a few local or isolated instances can
result in manganism in miners, lung cancer in chromate workers, nickel-carbonyl cancers in nickel workers, bony abnormalities from high fluoride wells and selenium poisoning
in seleniferous areas, a non-fatal ailment. Human exposures
have not only not increased to levels potentially of concern,
but in fact in some cases they have declined, because of
modern food practices in refining and processing our caloric
energy.

In order to ascertain recondite toxicity, it was
necessary to duplicate in small mammals the experiments modern
man is now unwittingly performing on himself during his life-
time (4,5). Therefore, a laboratory of wood covered with
plastic varnish was constructed on a remote Vermont hill-top,
designed to exclude contaminating metals from air and water.
Large numbers of rats and mice were exposed to single elements
in low doses for a life-time. Recondite toxicity was evaluated in terms of growth, life span, longevity, changes in serum,
tumors, microscopic pathology, disease, and tissue content of
the element fed. Reproduction of exposed animals was also
evaluated.

Under these life-time exposures, 30 elements were given
to mice and 20 to rats in drinking water. Little or no subtle
toxicity was shown by hexavalent chromium, fluoride, molybdenum, vanadium, nickel, arsenic, barium, aluminum, titanium,
zirconium, or niobium. Carcinogenesis was exhibited by

Table 1. Approximate Annual U.S. Industrial Consumption of Metals (1968), their Amounts in the Human Body and their Abundances on the Earth's Crust and in Sea Water.

Metal	Industrial Consumption (thousands of tons)	Amounts in Reference Man (mg)	Earth's Crust, ppm	Sea Water ppb	Disease from Excess
Essential for Life or Health					
Iron	109,000	4,200	50,000	10	Hemochromatosis
Calcium	86,273	1,000,000	36,300	400,000	
Sodium	15,091	100,000	28,300	10,500,000	
Potassium	3,230	140,000	25,900	380,000	
Manganese	1,050	12	1,000	2	Manganism
Zinc	1,278	2,300	65	10	
Copper	1,400	72	45	3	Hepatolenticular degeneration
Chromium	459	1.5	200	0.05	Cancer (CrVI)
Fluorine*	587	2,600	700	1,300	Fluorosis
Nickel†	170	10	80	5.4	Ni-carbonyl cancer
Magnesium	89	19,000	20,900	1,350,000	
Molybdenum	25	9	1	10	
Cobalt	6	1.5	23	0.27	
Vanadium	5	18	110	2	
Selenium	0.5	13	0.09	–	
Strontium	6	320	450	8,100	
Toxic to Living Things					
Lead	816	121 (9 to 480)	10	0.03	Plumbism

Element					Health effect
Antimony	19	8	0.2	–	Heart disease
Beryllium	0.3	0.04	2	–	Beryllosis
Cadmium	7	50/13	0.2	0.11	Hypertension, emphysema
Mercury	3	13	0.5	0.03	Poisoning
Slightly Toxic to Some Life Processes					
Tin	59	6	3	3	
Arsenic	22	18/+	2	3	Cancer
Tungsten	7	+	1	0.1	
Germanium	11	+	2	0.07	
Uranium	2.7	0.09	2	3	
Bismuth	1	0.2	0.2	0.017	Kidney disease, animals
Tellurium	0.1	8	0.002	–	
Palladium	0.02	+	0.01	–	
Rhodium	0.002	+	0.001	–	
Probably Inert in Living Things					
Aluminum	3,534	61	81,300	10	
Barium	700	22	400	30	Baritosis
Titanium	413	9	4,400	1	
Zirconium	61	420	160	0.022	
Lithium	2.6	2	30	180	
Silver	0.03	0.8	0.1	0.3	
Niobium	2	110	24	0.01	
Boron‡	70	14	3	4,600	

*Nonmetal, essential for healthy bones and teeth. + Essential for birds, possibly essential for mammals.
‡Essential for plants.
Numbers in italics are considered to be larger than normal for uncontaminated man.
Data from references 1-3.

Table 2. Effects of low Doses of Trace Elements Given to Mice in Drinking Water for Life.

Elements	Dose ppm	Males Median life span, days	Longevity days	Females Median life span, days	Longevity days	Remarks
Essential elements						
Fluoride	10	591	830	629	838	
Vanadate	5	500	779	590	805	
Vanadate+Chromium	5	569	804	615	856	
Chromium(III)	5	587	831	624	940†	
Chromium(VI)	10	493	721	570	830	
Nickel	5	479	896	703	929	Accumulates
Nickel+Chromium	5	528	830	591	842	-
Selenite	3	530	702	536	746*	Accumulates, amyloidosis
Selenate	3	528	672†	633	782*	Accumulates, amyloidosis
Non-Essential						
Toxic						
Lead	25	464*	865*	670	888*	Accumulates, infections
Lead+Chromium	25	547	850	632	804†	
Antimony	5	542	786	569	843	Accumulates
Cadmium	5	474*	814*	624	904	Accumulates
Mercury, CH₃	5	58*	-	65*	-	Accumulates
Slightly Toxic						
Tin	5	548	896	554	761	Accumulates
Arsenic	5	496	694*	548	789*	Accumulates
Germanium	5	478	712*	589	829	Accumulates

Tellurite	2	468	663	568	725*	Accumulates
Tellurate	2	551	662†	583	812	
Palladium	5	554†	815†	572†	851	Tumors
Rhodium	5	509	708	531	818	Tumors
Scandium	5	453	686	484	783	
Gallium	5	516	663	534	730*	
Yttrium	5	494	710†	574	908†	
Non-Toxic						
Titanium	5	511	770*	629	884*	Accumulates
Titanium+Chromium	5	558	832	568	760	
Zirconium	5	520	760	580	901	
Niobium	5	542	910†	536	803	Accumulates, fatty livers
Indium	5	487	678	504	785	

* Less than control value, $P < 0.05-0.005$
† Greater than control value, $P < 0.05-0.005$
Note: These groups are not comparable, except to their own controls, which are not shown. There were 54 mice in each group.

selenate, rhodium and palladium, but human exposures to
these elements are very low (Tables 2 and 3).

The 6 toxic metals will be considered separately, as
each one presents its own problems.

Lead: Lead is the largest contaminant of the environment. About one kilogram per day per capita is discharged
from the tail pipes of automobiles, to enter air, water and
soil (6). Human intakes from urban air are about equal to
absorptions from food. Lead accumulates in human bone and
other tissues with age. No lead was found in children's
bones, nor in the bones of Peruvian Indians circa 1200 A.D.,
and little in monk's bones up to 300 A.D. The body burden
of Reference Man today is 121 mg with a range of 9 to 480 mg.
It is probable that today's burden is 100 times that of
primitive man. The margin between "normal" blood levels
and levels considered to be toxic is becoming increasingly
narrow (7). At all levels of urban exposure, there is interference with erythrcyte delta-aminolevulinic acid dehydrase,
a red blood cell enzyme, in proportion to blood level (8).

Overt lead poisoning from children eating lead paint is
well known, with symptoms ranging from mental retardation to
convulsions and coma. Air borne lead from motor·vehicle
exhausts has been suspected to add to intakes of urban children playing at street level, causing poor performance in
school. Subclinical lead toxicity in urban dwellers has been
also suspected but not yet delineated. There is enough lead
(200 ppm) in the grass growing along a secondary highway in
our town to abort a cow; lead polluted hay killed a horse in
Wales and 13 horses in California (9).

In rats and mice exposed to lead in water, there was
early mortality, shortened life span and susceptibility to
infections. Old rats lost excessive weight. When chromium
was added to the water, however, effects on mortality and
life span were largely prevented. Lead given to rats
resulted in focal myocardial fibrosis in a quarter of the
animals.

The cause of this rapid build-up of environmental lead
is the use, since 1924, of alkyl lead added to gasoline as
an anti-knock agent, which has now contaminated the Northern
Hemisphere (10). One can predict with reasonable accuracy
that subtle lead poisoning under the guise of several vague

Table 3. Effects of Low Doses of Trace Elements Given to Rats in Drinking Water for Life

Element	Dose ppm	Males Median life span,days	Longevity days	Females Median life span,days	Longevity days	Remarks
Essential						
Vanadate	5	860	1147	961	1269	
Chromium(III)	5	922	1249	950	1288	
Nickel	5	837	1158	924	1182	
Selenite	3	58	-	348	-	Very toxic
Selenate	3	962	1117	1014	1184	Cancers, accumulates
Molybdate	50	916	1128	930	1270	
Toxic						
Lead	25	729*	1123*	727*	1162*	Accumulates, infections
Lead+Chromium	25	883	1071	-	-	Infections
Cadmium	5	822*	1156*	805*	1146*	Accumulates, hypertension
Antimony	5	766*	994*	805*	1092*	Accumulates, fibrosis, heart
Slightly Toxic						
Tin	5	876	1134	830	1160*	Fatty livers
Arsenic	5	825	1120	912	1249	Accumulates
Germanium	5	738*	1177	833	1231	Fatty livers
Tellurite	2	864	1163	908	1117	
Non-Toxic						
Zirconium	5	881	1127	947	1247	Fibrosis, heart
Niobium	5	892	1045*	998	1247	Fibrosis, heart

* Less than controls   †Greater than controls
Note: There were 52 rats in each group.

symptoms will soon appear in urban populations. The preven-
tion of this disease is the elimination of alkyl lead from
gasoline and to a minor extent, the abatement of lead fumes
from smelters and refineries. Once it has occurred, the
cure will be difficult and slow.

Cadmium: An insidious toxin, cadmium is a recent addition
to the growing list of environmental contaminants added to
low background levels. It is always associated with zinc
in nature, at a ratio of 1:500 or thereabouts. It has the
ability to displace zinc in biological systems, interfering
with zinc's function in enzymes.

   In rats fed low doses of cadmium for life, it accumulates in blood vessels, kidney and liver, and with this increase, zinc is also accumulated. High blood pressure
appears in increasing frequency after about a year (Table 4),
and is universal after 30 months of age (11). Associated
with hypertension are the typical scleroses of the renal
arterioles, enlargement of the heart and sclerosis of coronary
and other blood vessels (12). The degree of arteriosclerosis
in the aorta is enhanced (13). The findings in the rat reproduce the clinical and pathological picture of human hyperten-
sion.The cadmium in rat and human kidney and liver is bound
partly in metallothionein, a zinc and cadmium protein. In
blood vessels, it is probably bound to sulfydryl groups. Most
sulfur-containing chelating agents, such as glutathione, have
a higher affinity for cadmium than for zinc. The contrary
is true for most oxygen-nitrogen ligands (Table 5).

   There are a few chelating agents not containing sulfur,
however, with somewhat higher stability constants for cadmium
than for zinc. We chose one of these with high constants for
both metals, on the basis that cadmium is tightly bound in
tissues (glutathione, for example, may bind them less well
than do tissues). The zinc-sodium complex of cyclohexane
diamino tetraacetic acid (CDTA) when injected into rats,
reduced the blood pressure to normal and exchanged some renal
and hepatic cadmium for zinc. Blood pressures remained normal for many months (14).

   Injected cadmium also produced hypertension, relieved by
the zinc chelate (14). These observations have been confirmed in rats by Perry and Erlanger (15), who also fed cadmium
with resultant hypertension, and by Thind et al in the
rabbit and dog by injecting cadmium (16). Vascular reactivity

Table 4. Effects of Cadmium in Food and Water on Systolic
Blood Pressure of Rats.

Age, Months	0.1 ppm Cd B.P. Mm Hg	0.62 ppm Cd B.P. Mm Hg	5.1 ppm Cd B.P. Mm Hg	P*
Females				
3	85+2.2	109+3.2	–	< 0.01
4	87+4.4	110+3.7	–	< 0.01
5	81+2.2	112+6.1	–	< 0.01
7	–	115+4.0	–	< 0.01
12	84+5.8	–	211+8.3	< 0.01
13	82+3.4	–	–	–
17	92+4.9	–	182+12.6	< 0.01
24	84+3.8	–	205+ 10.9	< 0.01
30	99+4.2	–	229+12.9	< 0.01
Males				
12	106+5.7	–	124+5.6	< 0.025
17	94+3.8	–	122+4.5	< 0.01
24	79+3.6	–	137+6.2	< 0.01
30	93+5.1	–	198+7.9	< 0.01
Females,17				
Calcium in water			92†	
No calcium			253‡	< 0.001

* P is significance of difference between the 2 groups shown.
† 8 rats normotensive, 2 hypertensive (260 Mm Hg)
‡ All of 10 rats hypertensive
In all groups but three given calcium, there were 16-24 rats.
In the third column, 5.0 ppm cadmium was given in water.
Data from Kanisawa and Schroeder (12).

Table 5. Stability Constants of Some Chelating Agents for
Zinc, Cadmium and Mercury.

	Zinc	Cad- mium	Mer- cury	Differ- ence Zn–Cd	Anti- hyperten- sive
Desferri-Ferrioxa- mine B.............	11.07	7.88	–	-3.19	
Ferrocyanide.......	-15.39	-16.49	-11.95	-1.10	+
1,10-Phenanthro- lene..............	5.9	5.2	19.65	-0.7	
EDTA..............	16.50	16.46	22.15	-0.04	+
Thiocyanate........	2.17	2.24	17.47	0.07	+
Carboxypeptidase A.	10.5	10.8	21.0	0.3	
CDTA..............	18.67	19.23	24.30	0.56	+
BAL...............	13.48	?14.23	–	0.75	+
DTPA..............	18.17	18.93	26.27	0.76	
Azide.............	0.5	1.4	-9.15	0.9	+
2,2'-Thiobis(ethyl- aminodi(acetic acid)	13.17	14.0	23.81	1.83	+
Glutathione........	8.30	10.50	43.47	2.10	+
2,2'-ethylenedioxy- bis(ethylaminodi (acetic acid)......	14.5	16.73	23.20	2.23	

From Stability Constants. The Chemical Society, London
  1964.

was altered by cadmium, and was restored by the zinc chelate (17).

In respect to human hypertension, Perry and I found much greater amounts of cadmium in the urines of patients with hypertension than in normotensive urines (18). The death rate from hypertensive heart disease is highly correlated with the concentration of cadmium in air (19) and milk (20). People dying of hypertension had more cadmium in their kidneys and a higher ratio of cadmium to zinc than did people dying of accidents, coronary heart disease, cancer or miscellaneous causes from around the world (21). The prevalence of hypertension in hospital admissions was correlated with renal cadmium in various geographical areas of the world (22). Thind et al found higher serum levels of cadmium in hypertensives than in normotensives (23).

Not supporting this hypothesis are Morgan's findings of no differences in renal cadmium or zinc in a group of negroes with or without hypertension (24). Methodological differences may or may not account for these findings.

There is a curious paradox in the relation of cadmium to hypertension. Swedish cadmium workers exposed to dusts who accumulate amounts large enough to induce proteinuria and renal damage do not become hypertensive in excess. Likewise, persons exposed to large amounts in Japan with resultant decalcification of bone(itai-itai disease) do not have excess incidences of hypertension. Apparently low exposures induce hypertension; high exposures do not. When enough cadmium is accumulated to produce toxicity, hypertension is not especially common, whereas less than this amount induces hypertension. This phenomenon is unexplained.

In rats made hypertensive by partial constriction of a renal artery, the feeding of cadmium enhances the hypertension. When the operation fails to cause hypertension, the feeding of cadmium induces it (25). Thus, there is synergism between the two methods.

Cadmium in air appears to come mostly from industrial sources, not from the burning of fossil fuels (Table 6).

Cadmium has also been linked to pulmonary emphysema. Higher values have been found in liver and kidney of patients with emphysema and with lung cancer than in controls (26). Exposed

Table 6. Toxic Elements Entering Atmosphere from Fossil
Fuels and Ocean from Weathering.

Element	Coal ppm	Oil ppm	Air 1000 tons	Weathering Air 1000 tons	Air/total %
Cadmium	5	0.01	0.5+2.3*	0.5	84.8
Mercury	3.3	10	10	3.5	74.1
Bismuth	5.5	-	0.75	0.6	55.6
Lead	25	0.3	3.6+100†	131	44.3
Beryllium	3	0.0004	0.41	5.6	6.8
Germanium	5	0.001	0.7	12	5.5
Tin	2	0.01	0.28	11	2.5
Nickel	15	10	3.7	171	2.1
Arsenic	5	0.01	0.7	72	1.0
Sulfur	20,000	3,400	3,400	140,000	2.4

From: Bertine, K.K. and Goldberg, E.D. Fossil Fuel Combustion and the Major Sedimentary Cycle. Science 173: 233, 1971.

* From industry
† From gasoline additives

workers also can suffer from emphysema (27).

An isolated area of Japan was irrigated by waters of
the Jintsu River, into which a zinc mine and smelter was
pouring its effluents for many years. Growing grains and
fish were contaminated by cadmium and lead. Persons living
on those foods eventually exhibited severe softening of
their bones, with many fractures and deformities and much
pain (ouch-ouch disease). Their bodies contained large
amounts of cadmium and lead (28). No other examples of this
disorder, which was probably the result of low calcium levels
in the diet plus lead and cadmium, have been reported. This
disease represents an extreme example of multiple toxicity.

Nickel: Although nickel dusts are carcinogenic in exposed
workers, who get lung and nasal cancers (29), and nickel metal
can cause eczema (30), at present exposures in air and water
nickel offers little or no hazard to the general population.
When finely divided nickel is exposed to hot carbon monoxide,
as might occur during the incomplete combustion of coal and
petroleum (which contain nickel) the carbonyl can be formed.
Cigarette smoke contains nickel carbonyl, for example. This
compound is carcinogenic to rats and man (31). Life time
exposures of rats and mice to nickel acetate, however, did
not cause excess tumors or any other effect.

Antimony: Rats and mice exposed to antimony in water had
shortened life spans (32). In large doses, it will produce
heart disease. Present exposures are probably low, but this
metal is a potential hazard.

Beryllium: Air borne beryllium can cause berylliosis in exposed workers, and even in their families living near beryllium smelters and refineries (33). It is a serious, chronic
lung disease. Beryllium can cause cancer in rats. Although
general exposures are low, this metal ranks high in toxicity.

Mercury: There appears to be no hazard to the general population from inorganic mercury or aryl mercury compounds. Mercury is ubiquitous on this planet (34), and is found in every
living thing. Fish concentrate mercury and accumulate it
with age, and fish eaters, be they fish, flesh or fowl, have
more mercury than non-fish eaters. Mercury accumulates in
fish with age. This mercury is obviously non-toxic; if it
were, fish eaters would suffer from mercury poisoning, and
seals, otters, porpoises, and other mammals, including man,

living on fish would not survive.

Alkyl mercury is another matter. Alkylation of lead, tin and mercury converts mild to moderately toxic substances into compounds with high toxicity. There have been two outbreaks of methyl mercury poisoning, in Minimata and Niigata, Japan, resulting from dumping of methyl mercury catalysts by plastic plants into a bay and a river. This compound was taken up by fish and shellfish, and it poisoned 121 people in Minimata and 47 in Niigata, with 52 dead (35). When the dumping of methyl mercury was stopped, the epidemics died out, and no one else in Japan was affected.

Alkyl mercury was widely used as a fungicide on seed grains. When people disregarded the labels and ate treated wheat or other grains, they were poisoned by methyl or ethyl mercury. Such outbreaks have occurred from time to time in Iraq, India, and Guatamala, with many deaths. The most recent were in Almagordo, N.M. where a hog fed treated wheat was eaten; three people fell ill permanently (36). These poisonings from methyl mercury have little to do with the present scare of mercury in fish, for mercury has been in fish as long as there have been fish. Birds eating treated seeds, however, have died.

Analyses of fish, meat, eggs and poultry have shown that about 90 per cent of the mercury is in the form of methyl mercury, using the method of Westoo (37). This appears to hold for both fresh water and marine fish. Either the method of digestion or analysis gives falsely high values of methyl mercury, or methyl mercury in tissues is so tightly coordinated to sulfur and protein as not to be toxic.

Man has not polluted the oceans with mercury. If the world's annual consumption of mercury of some 10,000 tons were poured down the drain, only 3 tons would remain dissolved in 1.42 quintillion tons of sea water. Therefore, the mercury in marine fish is background mercury, and that in fresh water fish is background mercury, plus industrial mercury dumped in fairly stagnant lakes. In this case, fish take up extra mercury from the food chain.

Mercury in air comes largely from the burning of fossil fuels (Table 6).

Slightly Toxic Metals: There are nine metals exhibiting

slight or low grade toxicity to some forms of life. All of
them occur on the earth's crust in low concentrations, less
than 3 ppm, and in the body of man in less than 18 mg amounts.
Tin and arsenic are the two most highly consumed.

The toxicity of tin to mammals is low. Fed to mice and
rats for life, it did not affect life-span or longevity of
males, but lessened longevity significantly in female rats.
Tin accumulated in spleen but not in other organs appreciably
(38). No disorders dependent on tin are known in man.

Arsenic also had a low order of toxicity to rats, not
affecting growth, life span or longevity (38). It accumulated
to high concentrations in all tissues, especially aorta and
red blood cells. In mice, however, arsenic depressed life
spans and longevity significantly, accumulating in tissues
to less extent than in rats (39). In man, chronic arsenic
toxicity is accompanied by skin lesions which become cancer-
ous. There was an epidemic of arsenic toxicity and skin
cancers in Taiwan from naturally contaminated well-water
several years ago, and a child in Fallon, Nevada, also was
poisoned by arsenic in a well.

In preliminary experiments, tungsten given to rats
produced high blood sugars, probably as a metabolic antagon-
ist to molybdenum and chromium.

Germanium fed to mice shortened life-spans and longevities
of males, accumulated in tissues, especially spleen, and
suppressed the formation of tumors (39). In rats, it caused
fatty degeneration of the liver, shortened life spans and
suppression of tumors and accumulated in tissues to some
extent (38).

Palladium and rhodium were somewhat carcinogenic in mice
(40). Tellurium shortened life spans of female mice (41), but
not of rats (42). Human exposures to these elements are low.

Elements probably inert: There are eight elements which
are relatively inert on biological processes. Six of them
are abundant on the earth's crust and in man, and five are
consumed industrially in fair sized quantities.

Aluminum, barium and titanium accumulate in human lungs
with age, probably from natural dusts. Most of the barium
used industrially goes into muds for oil-well drilling.
Human exposures to these elements are wide-spread, and toxicity

except to certain compounds in large concentrations, are
rare. They can be neglected at present levels as hazards
to health.
Deficiencies of Essential Trace Elements.

Primary or secondary deficiencies of the bulk elements
are well known and need not be considered here. What are
less well understood are deficiencies of the essential trace
metals in human health. Although homeostatic mechanisms
are efficient in most cases, deficiencies can occur from
depletion of foods by refinement--such as flour, sugar and
fats or oils--and by marginal diets therefrom. We can
discern two deficiencies common in human beings in this
country, which may have adverse effects on health, and we
can suspect others from analytical data on human tissues
compared to data on other mammals.

Zinc: Low plasma levels of zinc have been described in
a variety of conditions and diseases (43); they may be
secondary to the disease. Slow wound healing, especially
of indolent ischemic ulcers of the lower extremities, has
been reversed by oral zinc supplements with often dramatic
results (44,45). Open operative wounds in young airmen have
shown rapid healing with oral zinc (46).

Circulation in ischemic extremities has improved with
oral zinc supplementation, relieving intermittent claudica-
tion (47). Low blood pressure in these extremities has
not improved, suggesting that zinc acts to dilate arteries
beyond an organic constriction. Even early gangrene has
healed under zinc supplements. These observations suggest
that marginal zinc deficiency may affect these conditions,
and that restoration of plasma zinc to normal helps to
correct them.

Chromium:Atherosclerosis is a disease common to all to
some extent and is conditioned by dietary factors. Biochemi-
cally it is characterized by elevated serum cholesterol levels
and diminished tolerance to glucose; anatomically by plaques
in the major arteries--aorta, heart, brain or legs. Rats
raised on low chromium intakes show glucose intolerance,
elevated serum cholesterol levels and plaques in their
aortas (48). These changes are duplicated when refined
sugar is the major source of calories. They are prevented
by feeding chromium in water, or by feeding dark brown sugar
which is fairly high in chromium.

The human body burden of chromium is low in American
subjects and high in foreigners, and declines with age in
the former. At various ages, from 10 to 25 percent of subjects are deficient. Persons dying of coronary heart disease had virtually no aortic chromium, compared to normal
amounts in those dying of other diseases (49). In some
persons, oral chromium complexes have lowered serum cholesterol and improved glucose tolerance, but the effects are
not consistent.

Other essential trace metals: School lunches were
found deficient or marginal in chromium, copper and manganese (50). Man contains much less manganese than do wild
mammals, and domestic animals raised for profit are fed
20-30 times the amounts in human diets. The reasons for
this discrepancy have not been described.
Carcinogenicity of Trace Metals.

Many metals are directly carcinogenic when injected into
animals, usually producing sarcomata (Table 7). By the oral
route, however, only four produce tumors: lead (at 0.1-1.0%
lead acetate in feed), selenium at 3 ppm in water, rhodium
and palladium at 5 ppm in water. In man, arsenic in large
doses can lead to skin cancer. By inhalation, three produce
lung cancers in animals, beryllium, hexavalent chromium,
and nickel, and the last two, lung cancers in heavily exposed
workers. Present levels in food, water and air appear to
offer no hazard to the population at large, except in certain
instances of high concentrations. We cannot exclude, however,
the possible effects of nickel carbonyl resulting from the
incomplete combustion of coal, petroleum and gasoline with
nickel additives.

SUMMARY AND CONCLUSIONS

The effects of various trace elements on human health
have been reviewed, in the light of annual industrial consumptions, natural abundances on the earth's crust and in
sea water, and contents of Reference Man, as well as changes
in experimental animals fed low doses for life.

Fourteen metals and two non-metals are essential for
life or health of mammals, all but two of which are mined
in large amounts, are abundant on the earth's crust, and
are found in human tissues in sizeable quantities. Most
of these metals are non-toxic to mammals in ordinary concentrations.

Table 7. Carcinogenicity of Metals in Experimental Animals.

Metal	Oral	Parenteral	Inhalation
Be	0	Osteosarcomas	Carcinomas,lung
Cd	0	Sarcomas, tera- tomas	-
Cr(VI)	0	Sarcoma	Carcinomas,lung (also man)
Co	0	Sarcomas	-
Fe-dex- trans	-	Sarcomas	-
Pb	Renal Carcinomas	Renal Carcino- mas	-
Ni	0	Sarcomas	Carcinomas,lung (also man)
Zn(chicks)	0	Teratomas(intra- testicular)	-
Se	Sarcomas, hepatomas	-	-
As	0 (carcinomas-man)	-	-
Ti	0	Carcinomas	-
Rh	lymphomas carcinomas	-	-
Pd	lymphomas carcinomas	-	-

From: Sunderman, F.W.Jr. Metal Carcinogenesis in Experimental Animals. Fd.Cosmet. Toxicology 9: 105-120, 1971,
and personal observations.

Five metals occur in low abundances, are consumed industrially in sizeable amounts, occur in the body of man and are toxic of themselves. Lead and cadmium are prevalent in man.

Nine metals have low orders of toxicity and occur at low crustal abundances. Eight have little or no toxicity, of which six are abundant.

Metals in the environment of potential hazard to man are lead, by far the largest pollutant, cadmium, which may influence hypertension and emphysema, nickel as the carbonyl which is carcinogenic, antimony which is toxic, beryllium which can cause beryllosis, and methyl mercury which is highly toxic. At present levels of exposure, lead, cadmium and possibly nickel are potentially hazardous to health. Only under special circumstances and in special compounds are a few other elements hazardous.

REFERENCES

1. 1967 Mineral Facts and Problems, US Dept of the Interior, Bureau of Mines, US Gov't Printing Office, Washington, D.C. 1970.

2. Tipton, I.H., personal communications.

3. Mason, B.: Principles of Geochemistry, 2nd Ed. Wiley: New York, 1952.

4. Schroeder, H.A., Vinton, W.H. Jr. and Balassa, J.J.: Effects of chromium, cadmium and other trace metals on the growth and survival of mice. J. Nutr. 80: 39, 1963.

5. Schroeder, H.A., Vinton, W.H.Jr., and Balassa, J.J.: Effects of chromium, cadmium and lead on the growth and survival of rats. J. Nutr. 80: 48, 1963.

6. Schroeder, H.A. and Tipton, I.H.: The human body burden of lead. Arch.Environ. Health 17:965, 1968.

7. Bryce-Smith, D.: Lead Pollution.-A growing hazard to public health. Chemistry in Britain 7: 54, 1971.

8. Ferm, V.H. and Carpenter, S.J.: The relationship of cadmium and zinc in experimental mammalian teratogenesis. Lab Invest. 18:429, 1968.

9. Rains, D.W.: Lead accumulated by Wild Oats (Avena fatua) in a contaminated area. Nature 233: 210, 1971.

10. Marozumi, M., Chow, T.J. and Patterson, C.: Chemical
composition of pollutant lead aerosols, terrestrial dusts,
and sea salts in Greenland and Antarctic snow strata.
Cosmochim.Acta 33: 1247, 1969.
11. Schroeder, H.A.: Cadmium hypertension in rats. Amer
J. Physiol. 207: 62, 1964.
12.Kanisawa, M. and Schroeder, H.A.: Renal arteriolar
changes in hypertensive rats given cadmium in drinking
water. J. Exp. Mol. Path. 10: 81, 1969.
13.Schroeder, H.A. and Balassa, J.J.: Influence of chromium,
cadmium, and lead on rat aortic lipids and circulating
cholesterol. Amer J. Physiol. 209: 433, 1965.
14.Schroeder, H.A. and Buckman, J.: Cadmium hypertension.
Its reversal in rats by a zinc chelate. Arch. Environ.
Health 14: 693, 1967.
15. Perry, H.M.Jr. and Erlanger, M.: Hypertension and tissue
metal levels after intraperitoneal cadmium, mercury and zinc.
Amer. J. Physiol. 220: 808, 1971.
16. Thind, G.S., Karreman, G., Stephan, K.F. and Blakemore,
W.S.: Vascular reactivity and mechanical properties of normal
and cadmium-hypertensive rabbits. J. Lab. & Clin. Med. 76:
560, 1970.
17. Schroeder, H.A., Baker, J.T., Hansen, N.M., Size, J.G.
and Wise, R.A.: Vascular reactivity of rats altered by cadmium
and a zinc chelate. Arch. Environ.Health 21: 609, 1970.
18. Perry, H.M.Jr. and Schroeder, H.A.: Concentration of trace
metals in urine of treated and untreated hypertensive patients
compared with normal subjects. J.Lab & Clin. Med.46: 936, 1955.
19. Carroll, R.E.: The relationship of cadmium in the air to
cardiovascular death rates. JAMA 198: 267, 1969.
20.Pinkerton, C., and Murthy, G.K.: Cadmium contents of milk
and cardiovascular disease mortality. Read before the Third
Annual Conference on Trace Substances in Environmental Health,
Columbia, Mo., University of Missouri, 1969, mimeographed.
21.Schroeder, H.A.: Cadmium as a factor in hypertension. J.
Chron. Dis. 18: 647, 1965.
22. Schroeder, H.A., Nason, A.P., Tipton, I.H. and Balassa,
J.J.: Essential trace metals in man: Zinc. Relation to
environmental cadmium. J. Chron. Dis. 20: 179, 1967.
23.Thind, G.S.: Role of cadmium in human and experimental
hypertension. J. Air. Poll. Control Assoc.(in press)
24.Morgan, J.M.: Tissue cadmium concentration in man.
Arch. Int. Med. 123: 405, 1969.
25. Schroeder, H.A., Nason, A.P., Prior, R.E., Reed, J.B. and
Haessler, W.T.: Influence of cadmium on renal ischemic hypertension in rats. Amer J. Physiol. 214: 469, 1968.

26. Morgan, J.M., Burch, H.B. and Watkins, J.B.: Tissue
cadmium and zinc content in emphysema and bronchogenic
carcinoma. J. Chron. Dis. 24:107, 1971.
27. Friberg, L.: Chronic cadmium poisoning. AMA Arch.Indus.
Health 20: 401, 1959.
28.Kobayashi, J.: Relation between the "Itai-Itai" disease
and the pollution of river water by cadmium from a mine.
Read before the 5th Internat. Water Poll. Res. Conf.,
Perganon Press Ltd., 1971.
29.Doll, R.: Cancer of the lung and nose in nickel workers.
Brit. J. Indust. Med. 15: 217, 1958.
30. Browning, E.: Nickel exxema. In: Toxicity of Industrial
Metals. 2nd Ed. Butterworth's: London, 1969.
31.Sunderman, F.W. and Donnelly, A.J.: Studies of nickel
carcinogenesis: Metastisizing pulmonary tumors in rats induced
by the inhalation of nickel carbonyl. Amer. J. Path. 46:
1027, 1965.
32.Schroeder, H.A., Mitchener, M. and Nason, A.P.: Zirconium,
niobium, antimony, vanadium and lead in rats: Life term
studies. J. Nutr. 100: 59, 1970.
33. Freiman, D.G. and Hardy, H.L.: Beryllium disease. The
relation of pulmonary pathology to clinical course and
prognosis based on a study of 130 cases from the US Beryllium
Case Registry. Human Pathology 1: 25, 1970.
34.Mercury in the Environment. Geol.Surv. Prof. Paper 713,
Dept of Interior, US Gov't Printing Office, Washington,
DC, 1970.
35.Takeuchi, T.: Biological reactions and pathological changes
of human beings and animals under the condition of organic
mercury contamination. Read before the Ecological Research
Branch, Internat. Conf. on Environ. Mercury Contamination,
Ann Arbor, Michigan, 1970, mimeographed.
36.Eyl, T.B.: Organic-Mercury food poisoning. New Eng. J. Med.
284: 706, 1971.
37. Westöö, G.: Determination of methyl mercury compounds in
foodstuffs. II. Determination of methylmercury in fish, egg,
meat and liver. Acta Chem. Scandinavica 21: 1790, 1967.
38. Schroeder, H.A., Kanisawa, M., Frost, D.V. and Mitchener,
M.: Germanium, tin and arsenic in rats. Effects on growth,
survival, pathological lesions and life span. J. Nutr. 96:
37, 1968.
39. Schroeder, H.A. and Balassa, J.J.: Arsenic, germanium,
tin and vanadium in mice. Effects on growth, survival and
tissue levels. J. Nutr. 92: 245, 1967.

40. Schroeder, H.A. and Mitchener, M.: Scandium, chromium (VI), gallium, yttrium, rhodium, palladium, indium in mice: Effects on growth and life span. J. Nutr. 101: 1431, 1971.

41. Schroeder, H.A. and Mitchener, M.: Selenium and tellurium in mice: Effects on growth, reproduction, survival and tumors. Arch. Environ. Health(in press)

42. Schroeder, H.A. and Mitchener, M.: Selenium and tellurium in rats: Effect on growth, survival and tumors. J. Nutr. 101: 1531, 1971.

43. Halsted, J.A. and Smith, J.C.: Plasma-zinc in health and disease. Lancet 1: 322, 1970.

44. Greaves, M.W. and Skillen, A.W.: Effects of long-continued ingestion of zinc sulfate in patients with venous leg ulceration. Lancet 11: 889, 1970.

45. Husain, S.L.: Oral zinc sulphate in leg ulcers. Lancet 11:1069, 1969.

46. Poires, W.J., Henzel, J.H., Rob, C.G. and Strain, W.H.: Acceleration of wound healing in man with zinc sulfate given by mouth. Lancet 1: 121, 1967.

47. Henzel, J.H., Lichti, E., Keitzer, F.W. and Deweese, M.S.: Efficacy of zinc medication as a therapeutic modality in atherosclerosis: Follow-up observations on patients medicated over long periods. Read before 4th Annual Conf. on Trace Substances in Environ. Health, University of Missouri, Columbia, Mo., 1970, mimeographed.

48. Schroeder, H.A., Mitchener, M. and Nason, A.P.: Influence of various sugars, chromium and other trace metals on serum cholesterol and glucose of rats. J. Nutr. 101: 247, 1971.

49. Schroeder, H.A., Nason, A.P. and Tipton, I.H.: Chromium deficiency as a factor in atherosclerosis. J. Chron. Dis. 23: 123, 1970.

50. Murphy, E.W., Page, L. and Watt, B.K.: Trace minerals in type A School lunches. J. Amer. Dietetic Assoc. 58: 115, 1971.

Supported by grants from the National Institutes of Health, HE 05076-12, CIBA Pharmaceutical Company and Cooper Laboratories, Inc.

Pesticides
in the Environment:
Recently Discovered
Analytical Problems

Edited by **Gunter Zweig**

INTRODUCTION

Dr. Gunter Zweig

Syracuse University Research Corporation

Syracuse, New York 13210

The purpose of this symposium is to high-light some of the recently discovered and partially solved problems in the analysis of trace amounts of pesticides and related compounds found in the environment. It is not the purpose of this symposium to review the methodology of pesticide residue analyses, because that has been the subject of many other symposia, books, articles, and so on. When this symposium was organized, the attempt was made to bring together a group of biologists, environmental experts and chemists, who would be able, first to illustrate the problems of sampling and detecting pesticide residues in the environment, and second to offer suggestions on the problems associated with the attempted identification of nanogram or picogram quantities of suspected chemicals. As for example, Compton's paper will cover the techniques of sampling the air for pesticide residues and identifying several unknown organic volatile components. Lincer's paper will discuss the occurrence of polychlorinated biphenyls in natural samples -- PCB's not being pesticide chemicals were included in this symposium, because for many years prior to the discovery of their presence in the environment, they were confused many times for DDT and related insecticides. Plimmer's paper describes the elegant techniques of gas-liquid chromatography wedded to mass spectrometry and using this tool in the identification of pesticides and degradation products. Bowman's presentation offers alternate solutions to solve problems of interference and identification by the use of selective

107

gas chromatographic detectors and distribution coefficients
between two immiscible solvents as tested by column
chromatography.

We hope that this published symposium will serve as a
guide to those researchers who may be concerned with the
most difficult task of identifying trace quantities of
organic pollutants in environmental samples.

POLYCHLORINATED BIPHENYLS: THEIR POTENTIAL INTERFERENCE
WITH PESTICIDE RESIDUE ANALYSIS AND PRESENT ANALYTICAL
STATUS

Jeffrey L. Lincer

Cornell University, Section of Ecology &
Systematics, Research Park, Bldg. 6, Ithaca,
N. Y. 14850

INTRODUCTION

The word "polychlorinated biphenyls (PCB's)" has
become common to analytical chemists working on pesticide
residue analysis. Although discussions of their properties
have been included in recent reviews (Peakall-Lincer, 1970,
Reynolds, 1971) a brief introduction to these properties
follows for those not familiar with this group of industrial
contaminants. The basic structure is depicted in Figure 1.
The positions marked by an x can be substituted by a
chlorine. Widmark (1968) calculated that of 210 possible
combinations, 102 are probable.

The following discussion will be restricted to those
PCB's which are or have been commercially available and
have been found in field-collected samples. In the
United States Aroclors (Monsanto Company trademark) are
designated by four digits (Monsanto Technical Bulletin
O/PL-306). The first two reflect the type of molecule:
12 - chlorinated biphenyls; 25 - a blend of chlorinated
bi- and triphenyls (75:25); 44 - a blend of chlorinated
bi- and triphenyls (60:40); 54 - chlorinated triphenyls.
The last two digits give the percent by weight of chlorine.
For example, Aroclor 1254, which is commonly found in field
samples, is a chlorinated biphenyl containing 54% chlorine.
It should be emphasized that we are dealing not with one
compound but a number of compounds. These are reflected
in a typical chromatogram (Fig. 2B from Reynolds, 1971).

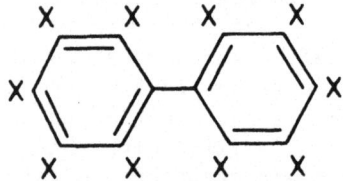

Figure 1.   Basic PCB structure

(X = Possible points for substitution of chlorine)

Bagley, Reichel & Cromartie (1970), carrying out mass
spectrometric work with Aroclor 1254 reported 18 compounds:
one containing three chlorine atoms; four containing four;
four containing five; five containing six; and four
containing seven. Note also (Figure 2) the coincidence
between the retention times of the commonly-found organochlorine pesticides and the peaks of Aroclor 1254.

    The physical and chemical properties of PCB's confer,
unfortunately, the appropriate qualities for persistence
in the environment and subsequent accumulation up food
chains (Table 1). These compounds are chemically inert.
They are not hydrolyzed by water and resist alkalies, acids
and corrosive chemicals. The viscous and more-highly-
chlorinated liquids and resins do not support combustion.
All Aroclor compounds are insoluble in water and glycerine
but vary over a gradient in their solubility in organic
solvents (Monsanto Technical Bull. O/PL-306). PCB's have
low volatility, boiling points ranging between 278 C and
415 C (Aroclors 1221 and 1268, respectively)(Penning, 1930).

    The suggested uses of PCB's are many and varied (Table
2). According to Monsanto (W. B. Papageorge - pers. comm.),
because of unintentional losses to the environment, the
only applications which now utilize PCB's are electrical
transformers, electrical capacitors and closed heat transfer
systems in non-food use when fire resistance is critical.
In addition, the use of highly-chlorinated triphenyl
Aroclors for extending the kill-life of organochlorine
pesticides has been suggested in the past by Monsanto and
considered by a number of investigators (Sullivan and
Hornstein, 1953; Hornstein and Sullivan, 1953; Tsao et al.,
1953; Duda, 1957; Lichtenstein et al., 1969). Although
these triphenyls have not yet been reported in environmental
samples, unlike the biphenyls they would not likely be
detected under the commonly-used operating gas chromatographic
conditions maintained for pesticide residue analyses
(Fig. 3 from Reynolds, 1971).

Potential Interference

    As Zweig (1963) mentioned, the pesticide analytical
chemist led a relatively serene life until about 1940. It
became more and more complex until around the early sixties
when the colorimetric methods ceased to give the kind of

Table 1. PCB residues in wildlife samples (mean ppm in extractable fat, range, and sample size). (Peakall and Lincer, 1970; data from Jensen et al., 1969).

	Baltic	Stockholm Archipelago
Mussell	4.3 (1.9 – 8.6) (40)*	5.2 (3.4 – 7.0) (15)
Herring	6.8 (0.5 – 23) (18)	5.1 (3.3 – 8.5) (4)
Seal	34 (16 – 44) (3)	30 (16 – 56) (3)
Guillemot eggs	250 (140 – 360) (9)	-----
White-tailed Eagle		
Pectoral Muscle	-----	14,000 (8,400 – 17,000) (4)
Brain	-----	910 (490 – 1,500) (3)
Eggs	-----	540 (250 – 800) (5)
Heron	-----	9,400 (1)

*Parentheses represent number of samples analyzed.

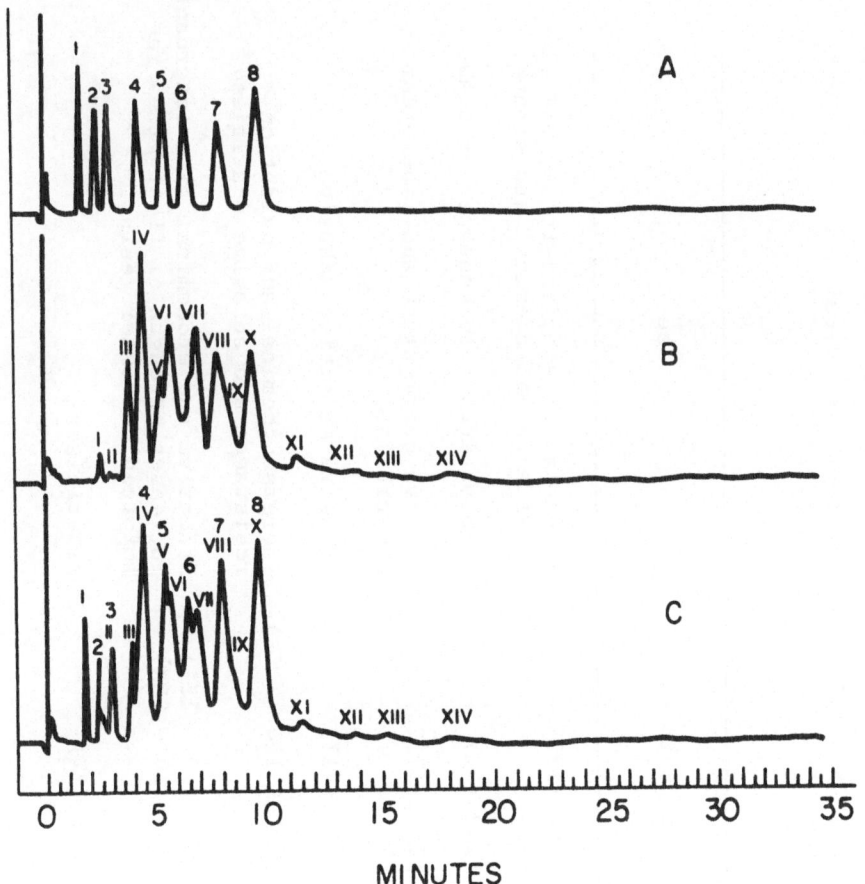

Figure 2.   PCB interference with organochlorine pesticide
            residue analysis on four percent SE-30/six percent
            QF-1 on 60/80 mesh Chromosorb W, 1/8" x 6' boro-
            silicate column. Chromatogram A: standard mixture
            of organochlorine pesticides; peak numbers:
            1 = 0.08 ng. of lindane, 2 = 0.10 ng. of heptachlor,
            3 = 0.10 ng. of aldrin, 4 = 0.14 ng. of heptachlor
            epoxide, 5 = 0.20 ng. of DDE, 6 = 0.20 ng. of
            dieldrin, 7 = 0.30 ng. of DDD, and 8 = 0.50 ng.
            of DDT. Chromatogram B: 5 ng. of Aroclor 1254
            (the 14 major peaks are numbered I to XIV).
            Chromatogram C: combination of above organochlorine
            standard pesticide mixture and Aroclor 1254.
            Injector 250°C., column 200°C., detector base
            250°C., N_2 at 20-30 cc./minute (Reynolds, 1971).

Table 2. Some uses of PCB's taken from Monsanto Technical
Bul. O/PL-306 (Peakall & Lincer, 1970)

Material with which Aroclor is combined	Aroclor used with wt %	Use
Polyvinyl chloride	Aroclor 1248, 1254 & 1260 (7 - 8%)	Secondary plasticizer to improve flame retardance and chemical resistance.
Nitrocellulose lacquers	Aroclor 1262 (7%)	Co-plasticizer to enhance resistance.
Polyvinyl acetate	Aroclor 1221, 1232, 1242 (11%)	Improve quick-track and fiber-tear properties.
Ethylene vinyl acetate	Aroclor 1254 (41%)	Pressure sensitive adhesive
Epoxy resins	Aroclor 1221 & 1248 (20%)	Increase chemical and oxidation resistance and adhesive qualities.
Polyester resins	Aroclor 1260 (10 - 15%) Aroclor 1260 (10 - 20%)	Effective and economical fire retardant Increases strength of fiber-glass reinforced polyester resins.
Polystyrene	Aroclor 1221 (2%)	Plasticizer

Table 2 – Continued

Material with which Aroclor is combined	Aroclor used with wt %	Use
Chlorinated rubber	Aroclor 1254 (5 – 10%)	Enhances resistance, flame retardance and improves electrical insulating properties.
Styrene–butadiene copolymer	Aroclor 1254 (8%)	Improves chemical resistance.
Neoprene	Aroclor 1268 (40%) Aroclor 1268 (1.5%)	Fire retardent. Injection moldings.
Crepe Rubber	Aroclor 1262 (5 – 50%)	Plasticizer in paint compositions.
Varnish	Aroclor 1260 (25% of oil)	Improves water and alkali resistance.
Wax	Aroclor 1242 (5%)	Moisture and flame resistance.

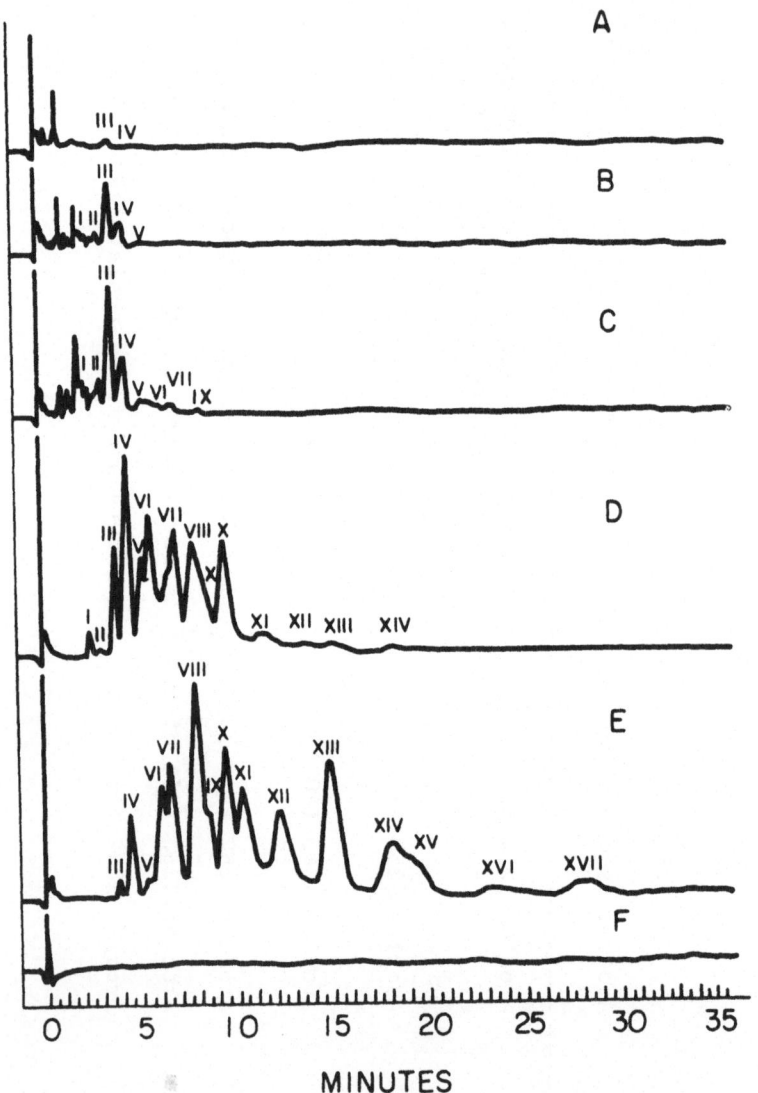

Figure 3.   GLC profiles of the more popular Aroclor
            mixtures under normal analytical conditions.
            Note: All peak numbers correspond to those of
            Aroclor 1254. Chomatograms: A = 5 ng. of
            Aroclor 1221, B = 5 ng. of Aroclor 1232, C = 5 ng.
            of Aroclor 1242, D = 5 ng. of Aroclor 1254,
            E = 5 ng. of Aroclor 1260, and F = 5 ng. of
            Aroclor 5460. GLC parameters as in Figure 2
            (Reynolds, 1971).

qualitative and quantitive data required of them thus
heralding the beginning of a new phase in pesticide
analysis.

Although residue chemists have been perplexed by
unknown gas chromatographic peaks for sometime, it appears
that Roburn in 1965 was the first to report them. Jensen
(Anonymous, 1966) first identified these unknown peaks
from museum feathers collected in Sweden as far back as
1944. Since then PCB residues have been reported in
wildlife from Canada (Holden and Marsden, 1967; Anderson
et al., 1969; Reynolds, 1971), Germany (Koeman et al.,
1967, Fiuczynski and Wendland, 1968), Great Britain
(Holden and Marsden, 1967; Holmes et al., 1967, Holden,
1970; Presst et al., 1970; Richardson et al., 1971), the
Netherlands (Koeman et al., 1967, 1969 a & b), Sweden
(Jensen et al., 1969) and the United States (Risebrough
et al., 1968; Risebrough, 1969, Anderson et al., 1969;
Risebrough, Florent and Berger, 1970; Henderson, Inglis
and Johnson, 1971; Rote and Murphy, 1971).

Henderson et al. (1971) provide us with an estimation
of PCB's in a variety of fish species throughout the
United States (Fig. 4). Although there appear to be
gradients in PCB levels, the variation in species and
trophic levels sampled negates further discussion of this
possibility.

Because of the possibility of interferences from a
variety of compounds including non-pesticide, electron-
capturing, polynuclear hydrocarbons (Abbott et al., 1966)
and "hybrid" residues (Bartha, 1969) confirmation of PCB
identification has been a necessity. Widmark (1967) reporting on the identity of unknown, but chlorine-
containing, compounds in field samples noted that most were
highly-chlorinated PCB's according to their mass spectra.
PCB's were found especially in fish and sea birds, but
also in conifer needles and human fat. Koeman, ten
Noever de Brauw and Vos (1969) reported on the mass
numbers and numbers of chlorine atoms per molecule of
unknown peaks from fish, mussels and seabirds. He concluded that most of these could be identified as PCB's.
Bailey, Bunyan and Fishwick (1970) confirmed their
presence as well as identifying a source of contamination.
They found PCB's (40% mixture) in imported cashew nuts
and their containers by mass spectrometry. Bagley et al.

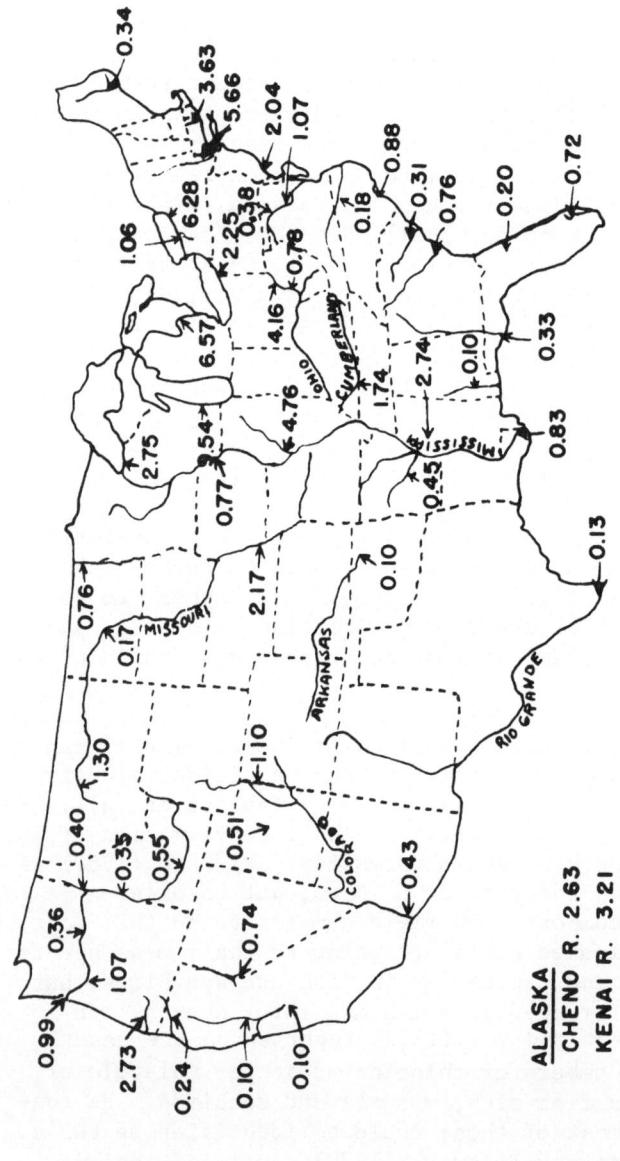

Figure 4. Estimated PCB levels of fish in United States (ppm based on wet weight, detection limit 0.10 ppm) (Peakall, in prep.; after Henderson, Inglis, and Johnson, 1971).

(1970) obtained mass spectra for 19 PCB's from two bald
eagle samples. Similarly Richardson et al. (1971)
confirmed the presence of PCB's in sea birds and fish oil
by combined gas-liquid chromatography-mass spectroscopy.
Bonelli (1971) using a limited mass plot in a similar
combination with the addition of a computer hook-up,
identified tetra- and hexachloro biphenyls in sewerage
and sturgeon ovary extracts, respectively.

In addition, work this year (Asai, et al., 1971)
indicates that carbon-skeleton chromatography may be a
useful technique for the qualitative differentiation
between PCB's and DDT and TDE.

Their ubiquity well reported and their presence confirmed, the next question becomes: What levels of interference can one expect from PCB's in the determination of
pesticide residue levels? Obviously, there is no one
answer to this question. PCB levels vary not only within
one geographic area (Risebrough et al., 1968; Jensen
et al., 1969) but there appears to be a basic difference
in the relative levels of PCB's between European and North
American samples (Richardson et al., 1971). That is, in
general in North American samples, DDT and metabolites
predominate over PCB's while at least in some European
samples the reverse is true. However, Zitko (1971)
reported PCB levels consistently in excess of DDE levels
in fish collected in New Brunswick and Nova Scotia.

A few studies to date have resulted in an estimation
of the potential error associated with ignoring the
presence of PCB's (Table 3). Several generalizations
can be drawn from these and other investigations carried
out particularly in North America: (1) In most cases,
the majority if not all the apparent DDE is real. This
is especially true with samples taken from (a) higher
trophic levels and (b) locations remote from PCB origin
where, for instance, PCB peak VI (at DDE, Fig. 2) is
relatively insignificant because of sample dilution to
quantify DDE or the relative decrease in PCB with distance
away from effluent nuclei (Risebrough et al., 1968);
(2) The majority of apparent TDE and DDT is in fact
contributed by PCB's; (3) The apparent heptachlor epoxide
and residue peaks occurring before it are probably
attributable to the respective pesticides. This is assumed
to be the result of differential diminution of early-

Table 3. Estimation of Error Contributed by PCB's When Separation Not Employed (expressed as percent of original sample due to PCB).

Species	Tissue	Sample Size (g)	p,p'-DDE	Chemical p,p'-DDE	p,p'-DDT	Heptachlor epoxide	Reference
Fish-eating birds[1]	egg	5	(2)	-57.7	-89.8	(5)	Anderson et al., 1969
California gull	fat	10	"	-72	-97		Vermeer & Reynolds, 1970
	ovary	"	"	-84	-84		"
	liver	"	"	-97	-87		"
	brain	"	"	-95	-96		"
	egg	"	"	-69	-100		"
Dbl. crested cormorant	egg	12(3)	"	-100	-75	+8	Reynolds, 1971
Mallard duck	egg	10(4)	"	-20	0	0	"
Common term	egg	2	"	-93	-30		"
Great blue heron	egg	8	"	-44	-44	-53	"

(1) Three double-crested cormorant and two white pelican eggs
(2) Authors suggested error was negligible
(3) Includes one pooled sample (10 eggs) and two single samples
(4) pooled
(5) Sample either had no original residue or was not checked for PCB contribution

arising PCB peaks (Koeman et al, 1969a, Grant, Phillips
and Villeneuve, 1971) and is especially true in samples
from raptorial and fish-eating birds. Work at Cornell
indicates that early-arising PCB peaks be diminished
approximately 70% in avian muscle and fat samples (Lincer
and Peakall, in prep.) (4) The PCB chromatogram profiles
which most commonly appear in field-collected samples
resemble Aroclor 1254 or 1260. (5) Generally, relatively
high levels of PCB occur where high levels of organochlorine pesticide residues occur but the converse is
not necessarily true. This is due to above-mentioned
trends within and between large geographic areas which
are reflections of local and regional land, PCB and
pesticide use as well as atmospheric conditions such as
air mass movements and water currents.

      The relative interference must, of course, be checked
for the particular operating condition used and/or
sample group being analyzed. Obviously, blanket dis-
missals of possible PCB interference based on other
investigators' results who may have employed different
extraction or clean-up techniques or different determinative
equipment on samples from various trophic levels, animal
or plant classes or geographic areas must be strenuously
avoided. The preceding five generalizations may not be
applicable to samples collected in Europe since (1)
little before and after separation residue quantitation
has been done outside of North America (2) PCB's appear
often to exceed organochlorine pesticide levels in
Europe and (3) PCB profiles often resemble other commercial
PCB mixtures like Clophen A 50 (Jensen et al., 1969)
or Phenoclor D P 6 (Koeman et al., 1967) which are not
necessarily identical to Aroclors 1254 or 1260. This last
aspect has particular importance to those of us working
on the biological effects of PCB's because the presence
of small amounts of more polar additives or contaminants
is apparently responsible for a significant increase in
toxicological effects (Vos and Koeman, 1970, Vos et al.,
1970).

Present Analytical Status

Separation

      The chemical techniques preliminary to quantitation
of residues in samples containing both PCB's and chlorinated

hydrocarbon pesticides fall into two groups -- those
necessitating the destruction or alteration of one or
more of the compounds and those which do not.

Included in the first group is nitration. Treatment
with a 1:1 mixture of sulfuric acid-nitric acid at 0 C
for five minutes destroys or alters aldrin, p,p'-DDE, p,p'-
DDD (TDE), p,p'-DDT and dieldrin such that they can no
longer be detected at the original position on the
chromatogram. This procedure leaves unaffected PCB,
lindane and BhC (Jensen and Widmark, 1967). A more
rigorous nitration with a 1:1 mixture of sulfuric acid-
fuming nitric acid for 15 minutes at room temperature
removes in addition to DDT and its related products,
aldrin, heptachlor, Kelthane, Perthane, Tedion, Telodrin
and Trithion while lindane, heptachlor epoxide, toxaphene
and Strobane are not removed (Erro, Bevenue and Beckman,
1967). Risebrough, Reiche and Olcott (1969) reported
that this nitration also removed the chromatographic
peaks of PCB's. Reynolds (1969) reported that his attempts
to nitrate samples were not fully successful in that there
appeared to be loss of some of the more volatile PCB
while peaks with longer retention times appeared. Armour
and Burke (1969) reported that complex chromatograms resulted after nitration which could not be related to the
unreacted DDT - PCB mixture, and nitration was not pursued
as a practical means of separating DDT and PCB for
further tests.

Saponification with alcoholic NaOH or KOH will de-
hyrodrochlorinate Perthane, Toxaphene, DDD and DDT to
their respective olefins (Archer and Crosby, 1966,
Klein and Watts, 1964) while Risebrough et al. (1969)
reported that PCB peaks are not removed or displaced but
gave no data.

The second and in some instances more desirable
group of analytical techniques allows the spacial separation of many chlorinated hydrocarbon pesticides from
PCB's. Reynolds (1969, 1970), reported on an activated
Florisil column technique which separated heptachlor,
aldrin, DDE and PCB with the first elution (60 ml n-
hexane) from lindane, heptachlor epoxide, DDD and DDT
with the second elution (40 ml 50 percent ethyl ether in
hexane). Armour and Burke (1970) developed a method
utilizing a silicic acid - Celite column eluting aldrin and

PCB with the first fraction (250 ml petroleum ether)
and lindane, heptachlor, heptachlor epoxide, dieldrin,
endrin, p,p'-DDE, o,p'-DDT, p,p'-DDT, and p,p'-DDD with
the second fraction (200 ml acetonitrile-hexane -
methylene chloride, 1:19:80). Koeman, Ten Noever de
Brauw and de Vos (1969a) using an activated Florisi
column eluted the apolar compounds including DDE and
PCB with hexane and then dieldrin and endrin with 10
percent diethyl ether. Snyder & Reinert (1971) using a
silica gel column and eluting with pentane (first
fraction) and benzene (second fraction) were able to
clearly separate Aroclors 1254 and 1260 from DDT, TDE and
DDE. Recovery values, generally exceeded 92%. Mulhern
(1968) reported on a method which utilized silica gel-
coated thin-layer plates and a hexane/ethyl ether (98:2)
solvent system. The plates were developed, sprayed with
a silver nitrate solution and exposed to U.V. light. The
plates were then divided into five horizontal sections.
Dieldrin, endrin, γ-BHC, heptachlor epoxide, p,p'-
DDD, p,p'-DDT, o,p'-DDT and p,p'-DDE (in that order)
were found in the first four fractions while most of
the interfering compounds found in wildlife samples were
present in the fifth zone. Bagley, Reichel and Cromartie
(1970) used this method but mentioned that zones three
and four contained practically all unknown components as
well as p,p'-DDT (zone three) and p,p'-DDE (zone four).
Armour and Burke (1969) used precoated (aluminum oxide)
sheets and n-heptane and 2 percent acetone/n-heptane for
solvent systems. PCB's (Aroclors 1254 and 1260) and DDE
were not separated but p,p'-DDT and p,p'-DDD were completely separated from PCB by both solvent systems. de
Vos and Peet (1971), using Kieselguhr G (impregnated
with paraffin) - covered glass plates and developing
three times in acetonitrile, acetone, methanol, water
(40:18:40:2) were successful at separating the components
of a PCB mixture (Phenchlor DP 6). They also suggested
that separation of PCB's from p,p'-DDT, p,p'-DDE, p,p'-
TDE, dieldrin, heptachlor epoxide and lindane was possible with their technique but no details were given.

Another possible means of separating interfering
substances is to use a series of differing polarity
columns in the gas chromatograph at the time of determination. This is less time-consuming and may be useful when
operating conditions can be selected such that PCB peaks
are absent in the region where sought pesticides emerge

(Simmons and Tatton, 1967).

Quantitation

Koeman, et al. (1969a) semiquantitatively measured
the residues in Japanese quail fed phenochlor DP6 by
using one of the peaks in a phenochlor DP6 mixture as a
standard. Risebrough (1969) quantitated relative levels
of PCB's by assuming that each PCB compound produced the
same peak height with the electron capture detector as
the same amount by weight of p,p'-DDE. After summing the
heights of the individual peaks, the total was multiplied
by a factor derived from measurements of standard
solutions with electron capture and microcoulometric
detectors. Jensen et al. (1969) reported PCB amounts as
the sum of all PCB components and based the estimate on
a combination of mass spectrometry and micro-coulometric
and electron capture detection. Even with this elaborate
approach these investigators suggest that the method
is still rough and may be correct only within a factor
of two. Anderson et al. (1969) devised a method for
obtaining a crude estimate of PCB residues as Aroclor 1254
from chromatograms where separation had not previously
been attempted. On an empirical basis, it was found
that Aroclor 1254 could be roughly quantitated by considering their peak 10 as p,p'-DDT and multiplying that
value by 10. Since this peak had originally been
quantitated as p,p'-DDT and in fact many of the original
samples contained little or no p,p'DDT it was possible to
estimate relative PCB values. Anderson and his coworkers,
after saponifying their samples to remove interfering
p,p'-DDT and p,p'-TDE, quantified PCB's as Aroclor 1254
by relating their sample peaks 9 and 10 to the corresponding
peaks of an Aroclor 1254 standard. Reynolds (1970)
employed a method similar to Koeman et al. (1969a) but
based quantitation on an average of two or more peaks.
In addition his results were reported as Aroclor 1254
or 1260 depending on the overall pattern of the chromatographic peak profile.

Most recently, Rote and Murphy (1971) reported on
a method, using GLC, which could be used for quantitation
of individual PCB isomers. Their method was based on the
fact that detector response is not the same for all PCB
components, varying with degree of chlorination. They
pointed out that if quantitation is based on the one

Aroclor, for instance, that the GLC profile resembles, a
figure would be arrived at which might be over <u>twice</u> that
resulting from the detector response curve approach.

Discussion

The analytical status of PCB's is no longer one of
infancy. By using standard extraction and clean-up
procedures appropriate for organochlorine pesticides
followed by close examination of resulting chromatograms
one can decide whether or not PCB-separation is required
(Reynolds, 1971). Depending on the pesticide, and
therefore separation, sought, very adequate separations
are available to the analytical chemist to separate
PCB's from pesticide residues (see analytical section).

As far as quantitation, there is not yet one standard
technique. It is clear that we are still relatively
unsophisticated in our PCB quantitation methodology and
will continue to estimate only relative amounts of PCB's
in field samples until we utilize an approach such as
that suggested by Route and Murphy (1971) which takes
into consideration the disparity in detector response
with varying degrees of chlorination. Hutzinger, Safe
and Zitko (1971) have made a major contribution in this
area by synthesizing and characterizing almost two dozen
individual chlorinated biphenyl. We should, however, not
lose sight of the fact that with reference to biological
significance, the correct order of magnitude and accurate
relative amounts of PCB give quite essential information
(Risebrough, pers, comm.). This is because most biological
effects are related to degree of chlorination and many
of the existing methods give a good indication of this
activity. Therefore, that information would be lost if
stress were placed only upon quantitating individual
peaks.

REFERENCES

Abbott, D. C., R. B. Harrison, J. O. G. Tatton and J.
    Thomson. 1966. Organochlorine pesticides in the
    atmosphere. Nature 211:259-261.

Anderson, D. W., J. J. Hickey, R. W. Risebrough, D. F.
    Hughes and R. E. Christensen. 1969. Significance
    of chlorinated hydrocarbon residues to breeding
    pelicans and cormorants. Can. Field Natur.
    83:89-112.

Anonymous, 1966. Report of a new chemical hazard.
    New Sci. 32:612.

Archer, T. E. and D. G. Crosby. 1966. Gas chromato-
    graphic measurement of toxaphene in milk, fat,
    blood, and alfalfa hay. Bull. Envir. Cont. and
    Toxicol. 1:70-75.

Armour, J. and J. Burke. 1969. Polychlorinated
    biphenyls as potential interference in pesticide
    residue analysis. F.D.A. Laboratory Information
    Bull. No. 918 11 pp.

Armour, J. and J. Burke. 1970 A method for separating
    polychlorinated biphenyls from DDT and its analogs.
    J. A. O. A. C. 53:761-768.

Asai, R. I., F. A. Gunther, W. E. Westlake and Y.
    Iwata. 1970. Differentiation of polychlorinated
    biphenyls from DDT by carbon-skeleton chromato-
    graphy. J. Agr. Food Chem. 19(2):346-398.

Bagley, G. E., W. L. Reichel, and E. Cromartie. 1970.
    Identification of polychlorinated biphenyls in
    two bald eagles by combined gas-liquid chomato-
    graphy-mass spectrometry. J. A. O. A. C. 53:251-261.

Bailey, S., P. J. Bunyan and F. B. Fishwick. 1970.
    Polychlorinated biphenyl residues. Chem. &
    Industry 30 May:705.

Bartha, R. 1969. Pesticide residue interaction
    creates hybrid residue. Science 166:1299-1300.

Bonelli, E. J. 1971. Computer-controlled GC/MS for
the analysis of polychlorinated biphenyls.
American Laboratory, February 8 pp.

de Vos, R. H. and E. W. Peet. 1971. Thin-layer
chromatography of polychlorinated biphenyls.
Bull. Envir. Cont. Toxicol. 6(2):164-170.

Duda, E. J. 1957. The use of chlorinated polyphenyls
to increase the effective insecticidal life of
lindane. J. Econ. Entomol. 50:218-219.

Erro, F., A. Bevenue and H. Beckman. 1967. A method
for the determination of toxaphene in the presence
of DDT. Bull. Envir. Cont. and Toxicol. 2:
372-380.

Fiuczynski, V. D. and V. Wendland. 1968. The population of _Milvus migrans_ in Berlin 1952-1967. J.
Ornithologie 109:462-471.

Grant, D. L., W. E. J. Phillips and D. C. Villeneuve.
1971. Metabolism of a polychlorinated biphenyl
(Aroclor 1254) mixture in the rat. Bull. Envir.
Cont. Toxicol. 6(2):102-112.

Henderson, C., A. Inglis and W. Johnson. 1971.
Organochlorine insecticide residues in Fish-Fall
1969 National Pesticide Monitoring Program.
Pestic. Monit. J. 5(1):1-11.

Holden, A. V. and K. Marsden. 1967. Organochlorine
residues in seals and porpoises. Nature 216:
1274-1276.

Holden, A. V. 1970. Source of polychlorinated biphenyl
contamination in the marine environment. Nature
228:1220-1221.

Holmes, D. C., J. H. Simmons, and J. O.'G. Tatton. 1967.
Chlorinated hydrocarbons in British wildlife.
Nature 216:227-229.

Hornstein, I. and W. N. Sullivan. 1953. The role of
chlorinated polyphenyls in improving lindane
residues. J. Econ. Entomol. 46:937-940.

Hutzinger, O., S. Safe and V. Zitko. 1971 Polychlorinated
    biphenyls: Synthesis of some individual chloro-
    biphenyls. Bull. Envir. Cont. Toxicol. 6(3):
    209-219.

Jensen, S. and G. Widmark. 1967. Organochlorine
    residues, OECD preliminary study 1966-67. Report
    given at the OECD pesticide conference.

Jensen, S., A. G. Johnels, S. Olsson, and G. Otterlind.
    1969. DDT and PCB in marine animals from Swedish
    waters. Nature 224:247-250.

Klein, A. K. and J. O. Watts. 1964. Separation and
    measurement of Perthane, DDD (TDE), and DDT in
    leafy vegetables by electron capture gas chromatography.
    J.A.O.A.C. 47:311-316.

Koeman, J. H., M.C. ten noever de Brauw and R. H. de
    Vos. 1969a. Chlorinated biphenyls in fish,
    mussels and birds from the River Rhine and the
    Netherlands coastal area. Nature 221:1126-1128.

Koeman, J. H., J. A. J. Vink and J. J. M. de Goeij.
    1969a. Causes of mortality in birds of prey and
    owls in the Netherlands in the winter of 1968-
    1969. Ardea 57:67-76.

Koeman, J. H., A. A. G. Oskamp, J. Veen, E. Brouwer, J.
    Rooth, P. Zwart, E. V. D. Broek and H. Van Genderen.
    1967. Insecticides as a factor in the mortality
    of the Scandinavian tern (Sterna sandvicensis.) A
    preliminary communication. Meded. Rijksfacultiet.
    Landbouwwetenschappen Gent. 32:841:853.

Lincer, J. L. and D. B. Peakall. (In prep.). Poly-
    chlorinated biphenyls: their pharmacodynamics and
    early peak diminution.

Lichtenstein, E. P., K. R. Schulz, T. W. Fuhremann, and
    T. T. Liang. 1969. Biological interaction between
    plasticizers and insecticides. J. Econ. Entomol.
    62:761-765.

Monsanto Technical Bulletins O/PL-311A, O/PL-306, and
    O-FF/1.

Mulhern, B. 1968. An improved method for the separation
    and removal of organochlorine insecticides from
    thin-layer plates. J. Chromatog. 34:556-558.

Peakall, D. B. (In prep.). Biological Magnification
    of Toxicants. Springer-Verlag.

Peakall, D. B. and J. L. Lincer. 1970. Polychlorinated
    biphenyls: Another long-life widespread chemical
    in the environment. BioScience 20(17):958-964.

Penning, C.H. 1930. Physical characteristics and
    commercial possibilities of chlorinated diphenyl.
    Ind. Eng. Chem. 22:1180-1182

Prestt, I., D. J. Jefferies, and N. W. Moore. 1970
    Polychlorinated biphenyls in wild birds in Britain
    and their avain toxicity. Environmental Pollution.
    1:3-26.

Reynolds, L. M. 1969. Polychlorobiphenyls (PCB's) and
    their interference with pesticide residue analysis.
    Bull. Envir. Cont. and Toxicol. 4:128-143

Reynolds, L. M. 1971. Pesticide residue analysis in
    the presence of polychlorobiphenyls (PCB's).
    Residue Reviews. 34:27-57.

Richardson, A., J. Robinson, A. N. Crabtree and M. K.
    Baldwin. 1971. Residues of polychlorobiphenyls
    in biological samples. Pestic. Monit. J. 4(4):
    169-176.

Risebrough, R. W., P. Reiche, D. B. Peakall, S. G. Herman
    and M. N. Kirven. 1968. Polychlorinated biphenyls
    in the global ecosystem. Nature 220:1098-1102.

Risebrough, R. W. 1969. Chlorinated hydrocarbons in the
    global ecosystem. In: Chemical Fallout. Ed. Berg,
    G. G. and M. W. Miller. Charles C. Thomas.
    Springfield, Ill., U.S.A. pp. 5-23.

Risebrough, R. W., R. Reiche, and H. S. Olcott. 1969.
    Current progress in the determination of poly-
    chlorinated biphenyls. Bull. Envir. Cont. and
    Toxicol. 4:192-201.

Risebrough, R. W., G. L. Florent and D. D. Berger. 1970.
    Organochlorine pollutants in peregrines and merlins
    migrating through Wisconsin. Can. Field Nat.
    84(3):247-253.

Roburn, J. 1965. A simple concentration-cell technique
    for determining small amounts of halide ions and
    its use in the determination of residues of
    organochlorine pesticides. Analyst 90:467-475.

Rote, J. W. and P. G. Murphy. 1971. A method for the
    quantitation of polychlorinated biphenyl (PCB)
    isomers. Bull. Envir. Cont. Toxicol. 6(4):
    377-384.

Simmons, J. H. and J. O'G Tatton. 1967. Improved gas
    chromatographic systems for determining organo-
    chlorine pesticide residues in wildlife. J.
    Chromatog. 27:253-255.

Snyder, D. and R. Reinert. 1971. Rapid separation of
    polychlorinated biphenyls from DDT and its
    analogues on silica gel. Bull. Envir. Cont.
    Toxicol. 6(5):385-390.

Sullivan, W. N. and I. Hornstein. 1953. Chlorinated
    polyphenyls to improve lindane residues. J.
    Econ. Entomol. 46:158-159.

Tsao, Ching-Hsi, W. N. Sullivan and I. Hornstein. 1953.
    A comparison of evaporation rates and toxicity to
    house flies of lindane and lindane-chlorinated
    polyphenyl deposits. J. Econ. Entomol. 46:882-
    884.

Vermeer, K. and L. M. Reynolds. 1970. Organochlorine
    residues in aquatic birds in the Canadian Prairie
    Provinces. Can. Field Nat. 84(2):117-130.

Vos, J. G. and J. H. Koeman. 1970. Comparative
    toxicologic study with polychlorinated biphenyls
    in chickens with special reference to porphyria,
    edema formation, liver necrosis and tissue residues.
    Toxicol. Appl. Pharm. 17:656-668.

Vos, J. G., J. H. Koeman, H. L. van der Maas, M. C.
    ten Noever de Brauw and R. H. de Vos. 1970.
    Identification and toxicological evaluation of
    two commercial polychlorinated biphenyls. Fd.
    Cosmet. Toxicol. 8:625-633.

Widmark, G. 1967. Possible interference by chlorinated
    biphenyls. J.A.O.A.C. 50:1069.

Widmark, G. 1968. Determination of the number of com-
    pounds which can result from the chlorination of
    biphenyl, and development of a simple system by
    which these may be codified. OECD report Sweden.

Zitko, V. 1971. Polychlorinated biphenyls and organo-
    chlorine pesticides in some freshwater and marine
    fishes. Bull. Envir. Cont. and Toxicol. 6(5):
    464-470.

Zweig, G. (ed.) 1963. Analytical methods for pesticides,
    plant growth regulators and food additives, vol.
    I. New York, Academic Press.

ANALYSIS OF PESTICIDES IN AIR

Bill Compton

Life Sciences Division, Syracuse University
Research Corporation, Syracuse, New York   13210

ENVIRONMENTAL DISTRIBUTION MECHANISMS OF PESTICIDES

The persistence of many chlorinated hydrocarbon pesticides governs to a great extent the modes by which worldwide distribution may take place: major source and related distribution mechanisms are listed in Table 1. Investigation of this table shows that there exists a relatively complex source-sink distribution system by which many persistent organo-chlorine and other types of pesticides may be re-circulated in the natural environment.

A primary source of environmental contamination is the application of pesticides by spraying and dusting. Spray particles decrease in size as they fall due to evaporation and co-distillation. If these particles evaporate completely before they reach ground level, they may remain suspended and are likely to drift[1]. If bodies of water are in the location of application, some pesticides will find their way directly into watercourses.

A second source, industrial manufacturing process losses, may introduce pesticides directly into air and water. A third source comes from food and vegetable product manufacturing processes. Pesticide residues are removed with the outside covering of the produce and discarded in large quantities. Subsequent decomposition of this refuse will introduce persistent pesticides into the environment[2].

TABLE 1

SOURCES AND DISTRIBUTION MECHANISMS OF ORGANIC

PESTICIDES IN THE ENVIRONMENT

1. Application by spraying and dusting.

2. Industrial manufacturing process losses.

3. Decomposition of pesticide-containing refuse.

4. Domestic use.

5. Entrainment of pesticide containing soils by wind and transport to other areas.

6. Volatilization of residues from soils and plant surfaces.

7. Vaporization from water surfaces (co-distillation processes).

8. Concentration in the surface organic film of the sea with enrichment in that layer and reinjection into the atmosphere by wave and wind action as aerosols.

Once the pesticides reach ground level several things may occur depending on their chemical characteristics, temperature, pressure, relative humidity and many other environmental factors. They may be adsorbed on soil particles, and due to entrainment may be translocated to other areas[3,4].

Volatilization of residues from soils and plant surfaces after application is known to occur. The gases will slowly dissipate into the ambient atmosphere, and either remain in the atmosphere as a gas or become adsorbed on existing suspended particulate matter.

Vaporization of pesticides from water surfaces is also thought to be a source since the loss of pesticides from static water surfaces has been demonstrated to occur[5].

Another proposed distribution mechanism is the possible concentration of pesticides in the organic surface film of the sea, thereby enriching that layer with pesticides. Wave and wind action subsequently reinject the pesticides as aerosols along with salt particulates[6]. Pesticides reach the ocean areas by air transport and are removed from the air by either gravitational fallout, or are washed out by precipitation[1]. This sink-source could be a considerable distribution mechanism, since it has been estimated that 24,200,000 million pounds of sea salt particulates are injected into the air annually[7].

The complicated sink-source relationships illustrated above indicates that a major mode of distribution of pesticides in the environment is transport by air. One would expect then, to be able to monitor the atmosphere and find persistent chlorinated pesticides present. This in fact has been done in Barbados[4]. The results of that study indicates the existence of a background level of chlorinated pesticides ranging from 0.006 to 0.38 picograms per cubic meter. The pesticides found at these concentrations were p,p'-DDT; o,p'-DDT; p,p'-DDE; p,p'-DDD; and Dieldrin.

MONITORING AIRBORNE PESTICIDES

Monitoring the ambient atmosphere for airborne pesticides has been conducted for many years, especially in areas of heavy pesticide usage. A multiplicity of sampling techniques exist, and may be categorized into four basic methods: (a) collection on an inert filter media, (b) collection in a liquid absorbing media, (c) collection on a solid adsorbent, and (d) collection using a combination of the above.

Collection on an inert filter media was used by Tabor[8] who analyzed samples collected by high volume air samplers which used a glass fiber filter mat to collect airborne particulate matter.

Liquid absorbing systems have been used extensively. Two typical systems used are those of Miles[9] whose system was composed of a series of modified Greensburg-Smith impingers, while Willis et al.[10] used a flask equipped with a gas dispersion tube. Both of these systems used ethylene glycol as the absorbing media. Bamesberger and Adams adapted the impinger technique in a unique way by developing a rotating disk impactor-midget impinger combination to collect gaseous and aerosol herbicides[11].

A dearth of information exists regarding the use of solid adsorbents alone as a method of airborne pesticide collection. The literature shows that solid adsorbents are generally used in combination with liquid absorbents. A florisil column was used by Yule[12], while Stanley[13] used alumina.

Although recovery studies for test pesticides have shown that the collection efficiency of each of the above methods is acceptable, each system is limited to a low sampling rate, from 0.1 cubic foot to 1 cubic foot per minute, and the absorbing media volume decreases rapidly after a few hours due to evaporation. At these low flow rates the volume of air sampled during the short time period sampling takes place does not provide enough suspended particulate matter to weigh accurately so that particulate concentrations may be calculated for correlation purposes.

In order to overcome these basic problems and others which evolve during routine sampling with liquid adsorbing systems, a monitoring unit was developed for field sampling of airborne pesticides. This unit was an outgrowth of a prototype field sampling instrument previously designed at the Syracuse University Research Corporation.

Since comparable units were not available commercially, the samplers were fabricated at SURC. A complete airborne pesticide collection system is shown in Figure 1.

FIGURE 1 Complete Airborne Pesticide Sampler

The shelter was a standard aluminum high volume air sampler shelter modified to provide added stability and rigidity. An aluminum plate was substituted for the sheet aluminum support plate supplied with the shelter. A second plate, identical to the first, was installed near the base of the seven-day skip timer, and attached to the legs of the shelter to give the unit added rigidity. Four aluminum angles were attached to the ends of the legs to provide adequate wind bracing.

The top plate formed the support for the air blower unit, flowmeter, and pesticide collection module, as well as the necessary piping. These parts are connected as shown in the schematic diagram in Figure 2. With the top of the shelter down and secure, air is drawn through the pesticide collection module by a high volume air sampling blower unit.

The collection module is composed of a Pyrex pipe 4" to 2" reducer assembled in such a manner that the two basic components, the glass fiber filter mat, located at the 4" end of the reducer, and the cottonseed oil-coated glass bead absorbing media in the 2" reducer end are held rigidly in place. The glass bead-cottonseed oil collection media is composed of 140 ml of 3 mm diameter smooth-surfaced Pyrex glass beads coated with 3 ml of refined cottonseed oil.

The air flow rate was adjusted to 10 CFM, as indicated on the flowmeter, by opening or closing the ½" gate valve, open to the atmosphere, and allowing the blower unit to free-wheel at its 40-60 CFM capacity while still pulling 10 CFM through the sampling module. This arrangement insures a longer running lifetime for the motor unit. The motor is turned on and off by the seven-day trip timer. Sampling is conducted for 24 consecutive hours.

The collection efficiency of this system has been determined for 14 vapor phase pesticides. The efficiencies are shown in Table 2.

An eight station network equipped with these monitors was operated for a 12-month period, from December, 1970 through November, 1971. During this period there were 26 scheduled sampling dates, with a potential number of 208 samples. Of this total, 202 samples were collected for a collection rate of over 97%, indicating that the instrument, when properly serviced and with a minimum of attention and

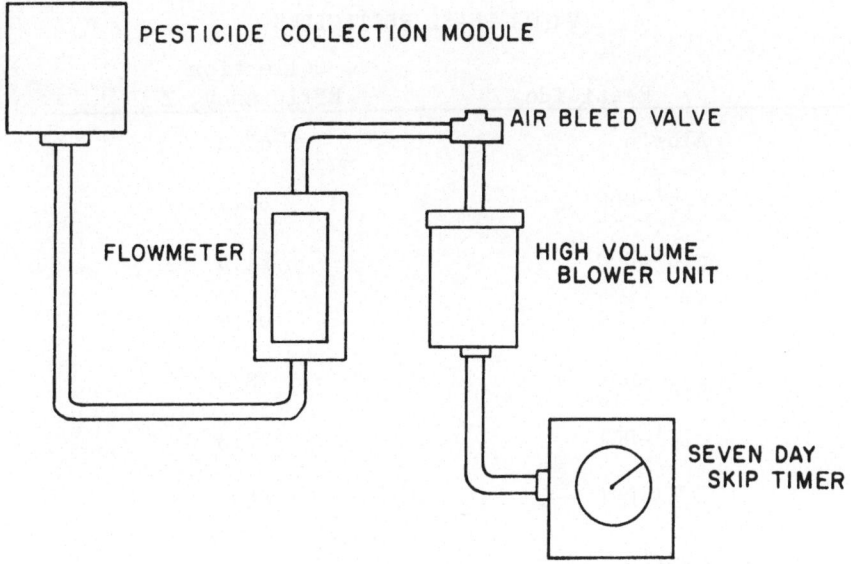

FIGURE 2   Schematic Diagram of Air Sampling System

Table 2

COLLECTION EFFICIENCIES FOR 14

VAPOR PHASE PESTICIDES

Pesticide	Collection Efficiency, %
Aldrin	67.2
o,p'-DDD	82.3
p,p'-DDD	91.5
o,p'-DDE	104.5
p,p'-DDE	78.8
o,p'-DDT	77.3
p,p'-DDT	88.5
Endrin	95.4
Heptachlor	70.6
Heptachlor epoxide	83.6
Malathion	70.5
Methyl parathion	78.8
Parathion	91.3
2,4-D Methyl Ester	100.5

maintenance, is highly reliable under varying ambient environmental circumstances.

EXTRACTION, SEPARATION AND CLEANUP OF
COLLECTED FIELD SAMPLES

When the pesticide collection module had been returned from the field, the component parts were immediately started through an extraction, separation and cleanup proceedure. The analytical scheme, outlined in Figure 3 was patterned after methods suggested in the Pesticide Analytical Manual Volume I[14] for fatty material.

The glass fiber filter, which had been previously tared, was removed from the module and reweighed after being placed in a desiccated drying chamber for 24 hours. The weight was recorded to the closest 0.1 mg.

The filter mat and cottonseed oil-coated beads were placed in a Soxhlet extraction apparatus. The glass reducer, screens, and spring were carefully rinsed with hexane three times. The rinses were placed in the extraction apparatus, and enough hexane added to bring the volume of solvent to 250 ml. The beads and filter were extracted for two hours at a cycle rate of at least once every 10 minutes.

The hexane extract was extracted with 50 ml. of 4% sodium bicarbonate and 35 ml. ethanol. The bicarbonate aqueous phase extraction was repeated two more times and the aqueous solutions pooled. This phase contained any 2,4-D acid that might have been present. The aqueous phase was extracted three times with 25 ml of chloroform. The chloroform washings were discarded, and the aqueous phase acidified with 25 ml of 10% sulfuric acid to free the 2,4-D acid. The acidified solution was extracted three times with 30 ml of chloroform each time. The pooled chloroform washings were drained into a 250 ml Phillips beaker, and evaporated carefully to dryness. The dry chloroform residues were then methylated with diazomethane to form the methyl ester of 2,4-D.

The original hexane extract was washed with water, dried by passing it through anhydrous sodium sulfate, and carried through an acetonitrile partitioning proceedure. The hexane solution was decreased slightly in volume, and carried

through a florisil column cleanup. All the chlorinated
and organophosphorus pesticides were recovered from the
florisil by elution from the column with 6% ether in
petroleum ether, and 50% ethyl ether in petroleum ether
fractions which were convenient for the study pesticide
mixture. The 6% eluate contained aldrin, o,p'-DDD, p,p'-DDD,
o,p'-DDE, p,p'-DDE, o,p'-DDT, p,p'-DDT, heptachlor, and
heptachlor epoxide.

The 50% eluate contained endrin, malathion, methyl
parathion, and parathion. No 15% ethyl ether in petroleum
ether eluate was collected since malathion divides between
the 15% and 50% eluate. As a result, malathion was eluted
directly from the florisil with the 50% ethyl ether-petroleum ether mixture.

The methyl ester of 2,4-D solution, and 6% and 50%
eluates were adjusted to 10 ml in anticipation of gas
liquid chromatography.

The average percent recovery of each pesticide carried
through the analytical scheme is given in Table 3.

GAS CHROMATOGRAPHIC ANALYSIS OF SAMPLE EXTRACTS

The analysis of the 2,4-D methyl esters, and the 6% and
50% ethyl ether-petroleum ether eluates were conducted by gas
liquid chromatography. The pesticide analyses were performed
on three Micro Tek (Tracor) Model MT-220 gas chromatographs.
Each unit was equipped with a Nickel-63, parallel plate,
electron capture detector operated in the DC mode. Two of the
units were equipped with high temperature flame photometric
detectors which were operated in the phosphorus mode.

These instruments are equipped with 6 ft by 4 mm glass
U-shaped columns. Four gas chromatographic column packings
were used during the course of the project. These were
(a) 1.5% OV-17/1.95% QF-1 on 80-100 mesh Supelcoport,
(b) 4% SE-30/6% QF-1 on 80-100 mesh Supelcoport, (c) 5%
OV-210 on 100-120 mesh Gas Chrom Q and (d) 10% DC-200/1.5%
QF-1 on 80-100 mesh Gas Chrom Q. Packing (a), (b), and (c)
were supplied by the Perrine Primate Laboratories, while
packing (d) was prepared using the mixed packing technique
described in Section 302.12b and Section 203.12a (1) in the
Pesticide Analytical Manual Volume 1. The gas chromatographic
operating conditions are given in Table 4.

Glass Beads Coated With Cottonseed Oil And Glass Fiber Filter

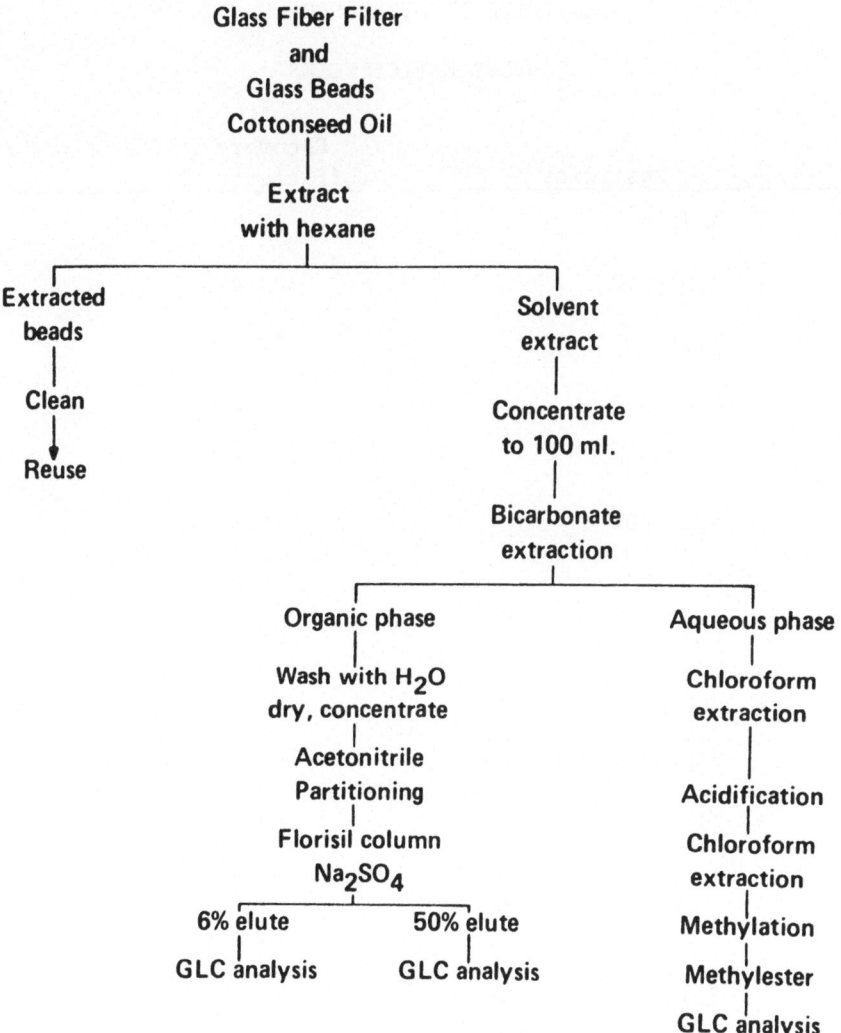

FIGURE 3   Analytical Scheme for Analysis

Table 3

PESTICIDE RECOVERY SCHEME

RECOVERY EFFICIENCIES

Pesticide	Recovery Efficiency, %
Aldrin	98.9
o,p'-DDD	107.4
p,p'-DDD	95.3
o,p'-DDE	94.2
p,p'-DDE	105.2
o,p'-DDT	95.5
p,p'-DDT	95.8
Endrin	92.3
Heptachlor	90.1
Heptachlor epoxide	108.4
Malathion	99.9
Methyl parathion	90.6
Parathion	84.6
2,4-D Methyl Ester	105.7

TABLE 4

Gas Chromatographic Operating Conditions

Column	1.5% OV-17/1.95% QF-1 on 80-100 mesh Supelcoport	5% OV-210 on 100-120 mesh Gas Chrom Q	4% SE-30/6% QF-1 on 80-100 mesh Supelcoport	10% DC-200/1.5% QF-1 on 80-100 mesh Gas Chrom Q
Column Temperature	200°C	180°C	200°C	200°C
Inlet Temperature	225°C	225°C	225°C	225°C
Detector Temperature, E.C.*	275°C	275°C	275°C	275°C
F.P.D.**	200°C	200°C	200°C	200°C
Bucking Range, Amperes E.C.	-2×10^{-8}	-2×10^{-8}	-2×10^{-8}	-2×10^{-8}
F.P.D.	$+2 \times 10^{-8}$	--	$+2 \times 10^{-8}$	$+2 \times 10^{-8}$
Input Attenuation	10^{2}	10^{2}	10^{2}	10^{2}
Output Attenuation	2 - 128	2 - 128	32 - 128	16
Polarizing Voltage	D.C. Mode	D.C. Mode	D.C. Mode	D.C. Mode
Carrier Gas	Prepurified Nitrogen	Prepurified Nitrogen	Prepurified Nitrogen	Prepurified Nitrogen
Inlet Pressure, psig	50	50	50	50
Flow Rate, Carrier Gas	60 ml/min	46 ml/min	60 ml/min	120 ml/min
Purge Flow Rate, E.C.	20 ml/min	34 ml/min	20 ml/min	--
F.P.D. Conditions:				
Hydrogen Flow	150 ml/min	150 ml/min	150 ml/min	150 ml/min
Oxygen Flow	20 ml/min	20 ml/min	20 ml/min	30 ml/min
Air Flow	20 ml/min	20 ml/min	20 ml/min	20 ml/min
Chart Speed	1/2 inch/min	1/2 inch/min	1/2 inch/min	1/2 inch/min

*Electron Capture Detector

**Flame Photometric Detector

Using the optimum operating parameters for a particular
gas chromatographic column-detector system, the sample
extracts were injected into the instrument. Choice of the
gas chromatographic column was predicted on the separation
of the study pesticides in the 6%, 50%, and methyl ester
2,4-D extracts.

Initial screening of the 6% eluates was done on a 1.5%
OV-17/1.95% QF-1 column using an electron capture detector.
This column separates all of the compounds expected to be
found in the 6% fraction with no overlapping peaks. For
confirmatory purposes either the 5% OV-210 column, or the
4% SE-30/QF-1 column with an electron detector was used.
An injection volume of 5 µl of the 10 ml eluate concentrate
was used.

Screening of the 50% eluate concentrate was done by
injecting 5 µl of the 10 ml eluate onto a 1.5% OV-17/1.95%
QF-1 column with an electron capture detector. This column
separates malathion, methyl parathion, parathion, and endrin
with no overlapping peaks. Confirmation of the organophosphorus pesticides was done by injecting 5 µl of the 50%
eluate which had been further concentrated to 1 ml, onto
either a 10% DC-200/1.5% QF-1 or 1.5% OV-17/1.95% QF-1
column with a flame photometric detector operated in the
phosphorus mode.

The 2,4-D methyl esters were screened by injecting
5 µl of the 10 ml extract onto a 5% OV-210 column. The
retention time of the methyl ester was 0.69 with respect
to Aldrin. Aldrin was injected as an internal standard
with each sample. Confirmation was done by injecting 5 µl
onto either a 1.5% OV-17/1.95% QF-1 column or 4% SE-30/QF-1
column.

Due to the nature of the sample extracts frequent
calibration checks were made to insure the sensitivity of
each detector system was maintained, and the column
performance remained at a high level of efficiency.
Therefore, 5 µl of standard which gave a response roughly
equivilent to the unknown was injected after each sample,
but not less frequently than every two injections. Standards
were injected at concentrations which gave a response in the
input and output attenuation settings used for the sample.

Suspected pesticides were identified by several techniques.

The most commonly used was to determine the retention
time of the unknown peaks and compare them with the
retention time of the standards run as calibration checks.
The other technique was to use one internal standard.
The suspected pesticides were then determined by comparing
their relative retention ratios with the internal standard.
This technique was used for all of the 2,4-D methyl ester
samples.

Quantitation was done by peak height for aldrin,
heptachlor, heptachlor epoxide, malathion, methyl parathion and parathion. The peaks exhibited by these pesticides
are sharp and lend themselves to this type of quantitation.

The limit of detection of each of the pesticides will
also establish the lower limit of equivalent concentration
in the air samples. Table 5 lists the detection limits of
the test pesticides by electron capture detection, in
nanograms. The lower limit of detection is defined as that
quantity of pesticide required to give a response on a
strip-chart recorder of 10% full scale deflection with a 1%
noise level.

The equivalent concentration in air of the test pesticides was calculated on the basis of injecting 5 microliters
of a 10 ml eluate volume. The 10 ml eluate represents those
pesticides collected from 403 cubic meters of air sampled.
Since this is a "floating" minimum equivalent concentration,
this value may be automatically lowered by reducing the
volume of the eluate.

CONCENTRATION OF AIRBORNE PESTICIDES IN THE FIELD SAMPLES

During the course of the field study, 202 samples were
analyzed for the study pesticides listed in Table 5.
Sampling was conducted in Winter Haven, Fla., Buffalo, Rome,
Jordan, Lafayette, Naples, and Syracuse, N. Y., and
Lubbock, Tex., respectively.

Pesticides were found on 17 sampling days of the 26
possible (65.4%). The concentration range of each pesticide
found at each sampling location is shown in Table 6. The
overall concentration ranges, in nanograms per cubic meter,
were aldrin, 0.22-2.15; DDE, 9.14-2.16; DDT, 0.54-1.01;
heptachlor, 0.09-1.32; malathion, 0.06-3.53; methyl
parathion, 0.02-0.84; parathion, 0.03-1.42; and total

Table 5

ESTIMATED DETECTION LIMITS OF AIRBORNE PESTICIDES,

NANOGRAMS/CUBIC METER*

Pesticide	Minimum Detectable Quantity, pg	Estimated Detection Limit, Ng/Cubic Meter
Aldrin	7	0.04
o,p'-DDD	24	0.12
p,p'-DDD	20	0.08
o,p'-DDE	30	0.15
p,p'-DDE	27	0.13
o,p'-DDT	21	0.10
p,p'-DDT	54	0.27
Endrin	58	0.29
Heptachlor	6	0.03
Heptachlor epoxide	15	0.07
Malathion**	247	0.09
Methyl Parathion**	42	0.04
Parathion**	89	0.05
2,4-D Methyl Ester	60	0.30

*Electron Capture Detection   ** Flame Photometric Detection

TABLE 6

Concentration Ranges of Test Pesticides in Collected Field Samples

Concentration Range, Nanograms/Cubic Meter

Pesticide	Winter Haven, Fla.	Buffalo, N.Y.	Rome, N.Y.	Jordan, N.Y.	Lafayette, N.Y.	Naples, N.Y.	Syracuse, N.Y.	Lubbock, Texas	Range, All Stations
Aldrin	1.04-1.08	0.22-1.96	--	2.15	--	--	--	--	0.22-2.15
DDE	0.27-1.94	0.37	--	0.14	0.15-1.52	0.63-2.16	0.38	0.31-1.36	0.14-2.16
DDT	2.03-5.24	1.74-1101	1.08	1.22-1.64	1.24-7.65	1.82-7.71	1.24-384	0.54-345	0.54-1101
Heptachlor	0.09-1.32	--	--	--	--	--	--	--	0.09-1.32
Malathion	0.11-3.53	0.07-1.81	0.08	0.08-0.22	0.12	0.06-0.17	0.10-0.11	0.06-0.61	0.06-3.53
Methyl Parathion	0.04	0.31	0.03	--	0.05	0.05	--	0.02-0.84	0.02-0.84
Parathion	0.07-0.58	0.03-1.42	--	0.12-0.27	0.24	0.13-0.35	--	0.07-1.03	0.03-1.42
Total 2,4-D	--	--	1.54	1.15	--	--	--	--	1.15-1.54

TABLE 7

AIRBORNE PESTICIDE DATA
VALUE BY DATE
NG/CM

STATION CODE: 33066001
STATION NAME: Buffalo, N. Y.

Date	Aldrin	DDD	DDT	Hepta-clor	DDE	Endrin	Hepta-clor Epoxide	Mala-thion	Methyl Para-thion	Para-thion	Total 2,4-D	Susp. Part. Matter
70/12/05	.00	.00	.00	.00	.00	.00	.00	.00	.00	.00	.00	84.3
70/12/14	.00	.00	.00	.00	.00	.00	.00	.00	.00	.00	.00	40.9
71/01/06	.00	.00	.00	.00	.00	.00	.00	.00	.00	.00	.00	97.5
71/01/13	.00	.00	.00	.00	.00	.00	.00	.00	.00	.00	.00	34.1
71/01/29	.00	.00	.00	.00	.00	.00	.00	.00	.00	.00	.00	711.9
71/02/11	.00	.00	.00	.00	.00	.00	.00	.00	.00	.00	.00	84.3
71/02/27	.00	.00	.00	.00	.00	.00	.00	.00	.00	.00	.00	68.3
71/03/08	.00	.00	.00	.00	.00	.00	.00	.43	.00	.00	.00	51.1
71/03/21	.00	.00	.00	.00	.00	.00	.00	.00	.00	.06	.00	60.6
71/04/04	.00	.00	.00	.00	.00	.00	.00	.00	.00	.00	.00	42.8
71/04/21	.00	.00	.00	.00	.00	.00	.00	.32	.31	.35	.00	109.2
71/05/04	.00	.00	.00	.00	.00	.00	.00	.00	.00	.00	.00	89.1
71/05/21	.22	.00	1.74	.00	.37	.00	.00	.00	.00	.00	.00	171.1
71/06/05	.00	.00	1101.38	.00	.00	.00	.00	.39	.00	.00	.00	74.1
71/06/14	.33	.00	3.71	.00	.00	.00	.00	.40	.00	.00	.00	71.1
71/07/01	.00	.00	3.03	.00	.00	.00	.00	.07	.00	.00	.00	82.6
71/07/16	.00	.00	.00	.00	.00	.00	.00	.55	.00	1.42	.00	71.1
71/07/26	.00	.00	.00	.00	.00	.00	.00	.00	.00	.00	.00	99.4
71/08/08	.00	.00	.00	.00	.00	.00	.00	.00	.00	.00	.00	72.1
71/08/25	.00	.00	.00	.00	.00	.00	.00	.00	.00	.03	.00	85.9
71/09/11	1.95	.00	.00	.00	.00	.00	.00	1.81	.00	.04	.00	280.0
71/09/21	.00	.00	.00	.00	.00	.00	.00	.00	.00	.00	.00	79.4
71/10/06	.00	.00	.00	.00	.00	.00	.00	.00	.00	.00	.00	41.3
71/10/23	.00	.00	.00	.00	.00	.00	.00	.00	.00	.00	.00	50.4
71/11/04	.00	.00	.00	.00	.00	.00	.00	.00	.00	.00	.00	68.4
71/11/17	.00	.00	.00	.00	.00	.00	.00	.00	.00	.00	.00	112.5

2,4-D acid, 1.15-1.54.

No DDD, endrin, or heptachlor epoxide were detected in any sample from any sampling site.

The data presented does not separate the isomers of DDE and DDT into the o,p'- or p,p'- isomers of each. The data for DDE represents p,p'-DDE only since o,p'-DDE was not detected in any sample. The primary DDT isomer was p,p'-DDT. The quantity of o,p'-DDT varied widely from sample to sample so that the ratio of o,p'-DDT to p,p'-DDT was not consistent.

The occurrence of the pesticides under study by date are shown in Table 7 for Buffalo.

REFERENCES

1. Mrak, E. M., "Report of the Secretary's Commission on Pesticides and their Relationship to Environmental Health", United States Department of Health, Education, and Welfare, Washington, D. C., December, 1969.

2. Westlake, W. E., and Gunther, F. A., "Occurrence and Mode of Introduction of Pesticides in the Environment", in Gould R. E. (ed.): Organic Pesticides in the Environment, Washington, D. C., Advances in Chemistry Series, American Chemical Society, 1966, pp. 110-121.

3. Risebrough, R. W., Huggett, R. J., Griffon, J. J., and Goldberg, E. D., "Pesticides: Transatlantic Movements in the Northeast Trades", Science, 159, 1233 (1968).

4. Cohen, J. M., and Pinkerton, C., Widespread Translocation of Pesticides by Air-Transport and Rain-Out", in Gould, R. F. (ed.): Organic Pesticides in the Environment, Washington, D. C. Advances in Chemistry Series, American Chemical Society, 1966, pp. 163-175.

5. Bowman, M. C., Acree, F., Lofgren, C. S., and Beroza, M., "Chlorinated Insecticides, Fate in Aqueous Suspensions Containing Mosquito Larva," Science, 194, 3650, 1480 (1964).

6.   National Academy of Sciences, "Chlorinated Hydrocarbons
     in the Marine Environment, A Report Prepared by the
     Panel on Monitoring Persistent Pesticides in the
     Marine Environment of the Committee on Oceanography",
     Washington, D. C., 1971.

7.   Study of Critical Environmental Problems (SCEP), 1970,
     p. 278, in National Academy of Sciences, "Chlorinated
     Hydrocarbons in the Marine Environment, A Report
     Prepared by the Panel on Monitoring Persistent Pesti-
     cides in the Marine Environment of the Committee on
     Oceanography", Washington, D. C., 1971.

8.   Tabor, E. C., "Pesticides in Urban Atmospheres",
     J.A.P.C.A., 15, 415 (1965).

9.   Miles, J. W., Fetzer, L. E., Pearce, G. W.,
     "Collection and Determination of Trace Quantities of
     Pesticides in Air", Environ. Sci. and Technol.,
     4, 420 (1970).

10. Willis, G. H., Parr, J. F., Papendick, R. I., Smith,
     S., "A System for Monitoring Atmospheric Concentra-
     tions of Field Applied Pesticides", Pesticide
     Monitoring Journal, 3, 172 (1969).

11. Bamesberger, W. L., and Adams, D. F., "Collection
     Techniques for Aerosol and Gaseous Herbicides",
     J. Ag. and Food Chem., 13, 552, (1965).

12. Yule, W. N., Cole, A. F. W., and Hoffman, I., "A
     Survey for Atmospheric Contamination following
     Spraying with Fenthion," Bull. of Environ. Contam.
     and Tox., 6, 289 (1971).

13. Stanley, C. W., Barney, J. E., Helton, M. R., and
     Yobs, A. R., "Measurement of Atmospheric Levels of
     Pesticides", Environ. Sci. and Technol., 5, 430 (1971).

14. Pesticide Analytical Manual Volume I, "Methods
     Which Detect Multiple Residues", Food and Drug
     Administration, U. S. Dept. of Health, Education,
     and Welfare, Washington, D. C. (1971).

THE SIGNIFICANCE OF PESTICIDE CONTAMINANTS

J. R. Plimmer

Plant Science Research Division, Agricultural
Research Service, U.S. Department of Agricultural
Beltsville, Maryland 20705

Pesticide manufacture normally yields a crude synthetic
product which can be purified or refined to a specified
content of the active ingredient. The major concern of
the analyst is to define the amount of active ingredient
present in formulations or batches of technical material
from the plant. Contaminants have rarely given cause
for concern, since biological testing with technical
material might be expected to reveal the presence of
impurities with significantly different properties from
the active compound. For example, the selectivity of
the herbicide amiben (3-amino-2,5-dichlorobenzoic acid),
is reduced by the presence of 2,5-dichloro-6-nitrobenzoic
acid or 6-amino-2,5-dichlorobenzoic acid and crops are
damaged by these compounds. Because these two compounds
possess undesirable biological activity, they must be
removed by rigorous control of the manfacturing process
(H. Segal, 1967).

It might be anticipated that the toxic properties of
a contaminant would be revealed by tests during the
development of a pesticide. However, the problem of
the chlorinated dioxins provides an example of such a
hazard which was not clearly defined for many years after
the introduction of pesticides potentially contaminated
by these compounds.

Although gross differences in biological activity
should become apparent during the early stages of

experimental testing of a pesticide, subtle effects due
to contaminants, such as synergism or antagonism, may
not be disclosed. Of importance are differences in
metabolic pathways which may lead to the accumulation of
a contaminant or metabolite derived therefrom. Ex-
ceptionally stable contaminants may resist metabolism
or degradation. Thus, while the pesticide is degraded,
a contaminant which is resistant to degradation may
accumulate through biological concentration and ultimately
present a hazard to human or animal life. The physical
properties of the contaminant: volatility, solubility,
polarity, etc., which determine its fate in the environment are not usually subject to the same scrutiny as
those of the parent compound.

The initial problem of the analyst is to identify
contaminants so that a systematic effort can be made to
determine the quantities present in a pesticide sample.
The question of significance arises when these answers
are known. Knowledge of the manufacturer's synthetic
route provides the most useful guide for the chemist
who wishes to identify contaminants. This information
is valuable, but the solution to the problem usually
entails lengthy isolation, purification and structural
determination procedures. The examples chosen show some
of the variety of techniques necessary for the solution
of these problems and provide some illustrations of the
significance of contaminants in a number of technical
pesticides.

The definition of technical standards for pesticides
has been confined primarily to the determination of the
amount of the active ingredient present in a sample.
In recent years, however, there have been a number of
developments which point to the importance of supplementing
this basic information. Methods of analysis such as
gas chromatography, capable of detecting picogram
quantities of many pesticides, are now available in the
majority of laboratories. There has resulted a rapid
increase in the volume of data generated concerning
quantities of pesticide residues in samples of all types.
The ecological implications of pesticide use have become
clearer as the results of monitoring programs accumulate.
A major problem in the interpretation of data relating
to the presence or the long-term effects of minute
quantities of pesticides or residues is to be absolutely

certain of the identity of compound under examination.
Gas, thin-layer or other chromatographic measurements
provide useful methods for quantitation. A single
column with a single detector, however, does not
provide sufficient evidence for positive identification.
Schechter (1968) has discussed this topic and the need
for suitable confirmatory techniques to obtain un-
equivocal proof of identity. At the present time, if
the problem of sensitivity can be overcome, the combination of gas chromatography-mass spectrometry appears to
be a most powerful and rapid method for many volatile
compounds. Cost factors, however, may limit its potential
development for wide-scale use.

The application of newer techniques for the analysis
of technical pesticides has revealed the identity of
many contaminants, but publications relating to this
information are rare in the literature. Admittedly
the nature of this information is somewhat ephemeral,
since there are variations in manufacturing processes
and there may be differences in production batches.
Much of the research reported in scientific journals
is performed with an intentionally purified pesticide
ingredient. While this is quite justifiable in the
laboratory, we need to feel confident that in practice
the findings can be extrapolated to the technical product
without major reservations.

Problems Due to Contaminants. Analytical Standards

Contamination is not necessarily present at the
time of manufacture. Conditions of storate may affect
the stability of a pesticide. Endrin, for example, is
reported to be unstable under certain conditions and
purified endrin standards must be protected from light
and heat. Ordinary storage at room temperature has been
found to reduce the endrin content of standards originally
more than 99% pure (Graham and Kenner, 1969). This
instability of endrin presents a greater problem for
storage of purified standards than for technical
preparations which may be stabilized by the presence of
impurities (Barlow, 1966).

Problems in Residue and Metabolism Studies

The following two cases from the literature have been

selected to exemplify the types of problem which have
been documented. The first relates to the presence of
a contaminant which was detected through residue studies.
The second is concerned with a controversy which may
hinge upon the presence of an impurity and its consequences
for metabolic studies

Gardner et al. (1969) reported that after kale was
sprayed with malathion (S- [1,2-bis(ethoxycarbonyl)ethyl]
0,0-dimethylphosphorodithioate) a residue could be
detected which was thought to be related to malathion.
However, the compound was found to be present in the
Malathion, 50% Emulsifiable Concentrate used for spray
preparation. It was identified as a homolog of malathion,
the ethyl butyl mercaptosuccinate S-ester with 0,0-
dimethylphosphorodithioate, by nuclear magnetic resonance
and mass spectrometry.

$$(CH_3O)_2PS.SCHCO_2R$$
$$|$$
$$CH_2CO_2R_1$$

Malathion $R=R_1=C_2H_5$

Homolog
$$\begin{cases} R_1=C_2H_5 \\ R=CH_2CH_2CH_2CH_3 \\ \text{or} \\ R_1=CH_2CH_2CH_2CH_3 \\ R=C_2H_5 \end{cases}$$

In some formulations examined, prepared in 1965 and
1968, the ratio of the homolog to malathion was about 1
to 5 as indicated by gas-liquid chromatography peak
areas. The oxygen analog of this compound was five
times as inhibitory toward bovine erythrocyte cholinesterase
as the oxygen analog of malathion. The homolog was also
detected as a residue on some samples of wheat and rice
containing malathion.

The synthesis of technical DDT gives a mixture of
products containing primarily p,p'-DDT [1,1,1-trichloro-
2,2-bis (p-chlorophenyl)ethane], some o,p-DDT [1,1,1-
trichloro-2- (p-chlorophenyl)-2-(o-chlorophenyl)ethane]

and other related impurities (Haller et al., 1945).
Fourteen compounds were isolated and identified in this
study which was accomplished by classical techniques of
organic chemistry. Nine kilograms of technical DDT was
used as starting material for the investigation. Studies
can be undertaken currently with much more sensitive
methods of analysis, but the fact must be borne in mind
that o,p-DDT is much more readily degraded than p,p'-
DDT and this difference may be reflected in rate studies
of technical DDT reactions.

There are several reports (Klein et al., 1964; Klein
et al., 1965; French and Jefferies, 1969; Ecobichon and
Saschenbrecker, 1968) of the conversion of o,p'-DDT to
p,p'-DDT in mammals or birds. It is difficult to
postulate a plausible biological mechanism for this
conversion and the findings have been challenged by
Bitman et al. (1971). They proposed that supposedly pure
samples of o,p'-DDT contained sufficient p,p'-DDT to
account for the appearance of p,p'-DDT as a metabolite.
They were unable to repeat the conversion with
radioactively labeled o,p'-DDT and they demonstrated
that the p,p'-DDT content of some samples of "pure"
o,p'-DDT was over 1% in some cases. Accumulation of p,p'-
DDT in tissue occurs since o,p'-DDT is rapidly converted
to hydroxy and methoxy metabolites which are excreted.
p,p'-DDT is not metabolized and is stored in the lipid
of the animal body. Thus measurements over a period of
time show that when relatively pure o,p'-DDT is fed, the
major component o,p'-DDT disappears with a simultaneous
appearance of the contaminant p,p'-DDT. The contaminant
p,p'-DDT was not detected in 99% o,p'-DDT samples
containing (1 µg/ml) analyzed by gas chromatography but
could be detected in concentrated (100 µg/ml) solution.

The Dioxin Problem

The chlorinated dibenzo-p-dioxins and some related
neutral compounds are contaminant by-products which may
arise during the manufacture of chlorinated phenols. These
compounds are extremely toxic and their presence in
chlorophenols or pesticides, synthesized from chlorophenols, has presented serious problems. These have
occurred during the manufacture of the herbicide 2,4,5-
trichlorophenoxyacetic acid (2,4,5-T) and in the preparation of commercial fatty acids for chicken feeds. On

several occasions outbreaks of chick edema disease have
occurred and these have been ascribed to the presence of
dioxin contaminants in the feed fats.

2,4,5-T Herbicide

The herbicide 2,4,5-T is used to control brush,
woody plants, and herbaceous broadleaf weeds. It was
first registered in March 1948 by the Amchem Products
Company of Ambler, Pennsylvania. It has been estimated
that 3.4 million acres of farmland and 4.5 million
acres of non-farmland were treated in 1969 with 8.9 million
pounds of 2,4,5-T. 2,4,5-T is classed as a plant hormone
but its mode of action is not yet fully elucidated. The
acute toxicity (LD_{50}) of 2,4,5-T ranges from 300 to
1,000 mg/kg depending on the particular animal species
(Anon 1971a, b). The finding "that 2,4,5-trichloro-
phenoxyacetic acid is teratogenic and fetocidal in two
strains of mice when administered either subcutaneously
or orally and in one strain of rats when administered orally
and that the incidences of both cystic kidney and cleft
palate were increased in mice . . ." (Courtney et al.,
1970) provoked controversy over the continued use of the
herbicide. However, it was later stated that the sample
of 2,4,5-T used was contaminated with 27 ± 8 ppm of
2,3,7,8-tetrachlorodibenzo-p-dioxin (TCDD). The actions
which followed this disclosure served to focus attention
on the problem due to the dioxin contaminants.

The problem has existed for some time. In 1949
there was an accident in which intermediate chemicals in
2,4,5-T manufacture were released into a plant owned by
the Monsanto Chemical Company. One hundred and seventeen
workers were affected and displayed symptoms of
"chloroacne." The severe skin disease, recognized since
1899 (Herxheimer, 1899) results from exposure to chlorinated
aromatic compounds and is characterized by areas of
blackheads, comedones, pustules and inclusion cysts over
the neck, back and chest. Other symptoms have also been
described and symptoms may be present for several years.
Similar incidents occurred in plants manufacturing
2,4,5-T and 2,4,5-trichlorophenol in Germany and the
responsible agent was found to be TCDD (Kimmig and
Schulz, 1957). The synthesis and biological examination
of TCDD demonstrated the extremely toxic nature of this
compound and its effectiveness in producing chloroacne.

More recently, in response to the increased demand for
2,4,5-T the Dow Chemical Company changed its method of
production in 1964 and, at the same time, 60 workers
were found to be affected with chloroacne. Investigations revealed the influence of temperature and pressure
on the amount of dioxin formed during the synthesis of
the intermediate 2,4,5-trichlorophenol. Since this
occurrence, a major research effort has been devoted to
the examination of many aspects of the dioxin problem
(Anon. 1970).

     2,4,5-T is manufactured by the condensation of
sodium 2,4,5-trichlorophenoxide (1) with sodium chloroacetate under mildly alkaline conditions:

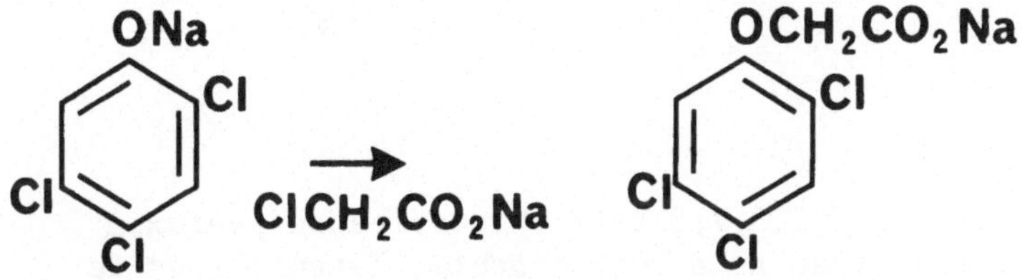

The free acid is precipitated by acidification of the
sodium salt (2) with sulfuric acid. The starting material,
sodium 2,4,5-trichlorophenoxide, is made by the hydrolysis
of tetrachlorobenzene (3):

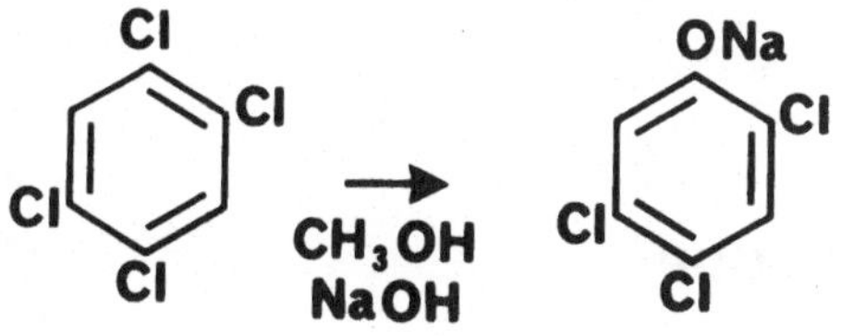

The reaction is conducted at a high temperature under pressure. An intermediate in the reaction is presumed to be 2,4,5-trichloroanisole. Rigorous control of temperature and pressure is necessary to obtain optimum yield and minimize side reactions. If the temperature is not carefully controlled the reaction:

(1)   (4)

may take place and 2,3,7,8-tetrachlorodibenzo-p-dioxin (TCDD) (4) is formed as a contaminant. Technical grade 2,4,5,-T contains 90-92% 2,4,5-trichlorophenoxyacetic acid and 8-10% impurities. 2,4,5-Trichloroanisole as an intermediate may remain unreacted, or it may be hydrolyzed under the synthetic conditions to give 5-methoxy-2,4-dichlorophenoxyacetic acid (5) or 2-methoxy-4,5-dichlorophenoxyacetic acid (6) by reaction with sodium chloroacetate:

(5)   (6)

Since technical chloroacetic acid normally contains some dichloroacetic acid, bis (2,4,5-trichlorophenoxy) acid (7) is formed.

Standard analytical procedures for 2,4,5-T involve a titration to determine acid content. Although the product commonly marketed has an acid content 97-98%, this analysis does not reflect the presence of several impurities (Table 1) and specific gas chromatographic analysis is necessary to determine actual 2,4,5-trichlorophenoxyacetic acid content.

TCDD is the most toxic impurity present in 2,4,5-T. Therefore efforts towards its analysis, control and elimination have received a good deal of attention since

Table 1.   2,4,5-T Product Composition (Anon. 1971a)

% by weight

91.0±1.0	2,4,5-trichlorophenoxyacetic acid
0.5±0.1	2,4,5-trichloroanisole
2.0±0.5	5-methoxy-2,4-dichlorophenoxyacetic acid
2.0±0.5	2-methoxy-4,5-dichlorophenoxyacetic acid
0.3±0.1	2,4,5-trichlorophenol
3.0±0.5	bis-2,4,5-trichlorophenoxyacetic acid
0.2±0.1	SO_4
0.2±0.1	Na salt of 2,4,5-T
0.5±0.2	H_2O
less than 1 ppm	TCDD

The above table represents the product of one manufacturer
and the content of particular samples can vary from
batch to batch and may also vary among producers.

the problem was brought to light. There are several
procedures for the analysis of TCDD in technical
phenoxyacid herbicides or their salts. A typical method
involves treatment of the sample with methanolic
potassium hydroxide followed by hexane extraction.
The washed and dried hexane extract is eluted from an
alumina column by petroleum ether followed by 5% diethyl
ether in petroleum ether. The dioxin content of the
latter eluant is measured by electron capture (Ni^{63} detector)
gas chromatography after volatilizing the diethyl ether
and adjusting the volume with hexane (Woolson and Thomas,
1971).

The problem of dioxin contamination is not restricted
to 2,4,5-T. The term, "chick edema factor", was used
to describe a complex mixture of related compounds
responsible for the disease in chickens known as chick
edema. This disease is characterized by hydropericardium
and liver and kidney damage. The first outbreak in the U.S.A.
caused the death of millions of broiler chicks and
sporadic outbreaks have occurred since that time.
The cause was traced to contaminated lots of edible
tallow used for production of food grade fatty acids
(Friedman et al., 1959).

Gas-chromatographic analysis and biological assay
were used to estimate the factor quantitatively on an
empirical basis, but its chemical nature proved difficult to determine. Several years of research were
required before, research workers at the Procter and
Gamble Co. showed that a crystalline factor, isolated from
an animal feed fat, and which could produce hydropericardium in chicks, had the structure 1,2,3,7,8,9-
hexachlorodibenzo-p-dioxin (Cantrell et al., 1967). The
structure was determined by X-ray crystallography
(Cantrell et al., 1969). An earlier report from the same
laboratory had described the isolation of two compounds
from toxic fat (Wootton and Courchene, 1964). Based on
mass spectral evidence the formula $C_{14}H_{10}Cl_6$ had (wrongly)
been allocated to these two compounds and the investigators
concluded that they were isomeric chlorinated perhydrophenanthrenes. Approximately 20 µg was used for the
mass spectrum which was scanned in 3 seconds, since
fragmentation patterns appeared to change with time.
The misinterpretation of this spectral data followed
from the possible ambiguity in interpretation of the

fragment patterns at low resolution. A loss of 63
mass units was interpreted as the loss of C_2H_4Cl; whereas,
in fact, this corresponds to the loss of $COCl$. The
relative intensities of the peaks at m/e 388, 390,
392, 394, 396 and 398 were those calculated for 6 Cl
atoms on the basis of the presence of the isotopes ^{35}Cl and
^{37}Cl in the ratio 100:32.7. Fragmentation pathways
involving chlorine loss could be clearly identified but
the fragment of 28 mass units could reasonably be assigned
the composition CO or C_2H_4. The dioxins behaved in a
manner similar to that expected for perhydrophenanthrenes
and comparison of the retention times of the perhydrophenanthrenes with those of the product of hydrogenation
of the toxic factor on a silicone-coated Chromosorb
W column provided evidence in support of this hypothesis.

It is instructive to examine the literature in
retrospect and point to significant data as they emerge.
For example, the chick edema factor was originally
thought to possess a polynuclear aromatic or modified
steroidal structure. The recognition that it contained
chlorine was a significant factor in its classification
(Tishler, 1960). The similarity of the absorption
spectrum of the toxic factor to those of the chlorinated
napthalenes provided a lead, since pentachloronapthalene
was known to cause hyperkeratosis in cattle and
chickens (Yartzoff et al., 1961). However, chlorinated
naphthalenes that produced hyperkeratosis in cattle
proved to be inactive in chick edema test. Since the
reports of Cantrell's work in 1967, it has been well
established that mixtures of chlorinated dioxins are
responsible for the symptoms due to toxic fats.

In view of the uncertainty surrounding the nature
of the chick edema factor and its complexity, the Federal
Register (1961) prescribed standards for food additive
fatty acid, based on freedom from the toxic factor by
bioassay, or by the absence of chromatographic peaks
with a retention time relative to aldrin between 10 and
25 using the electron-capture gas chromatographic
method (Neal, 1967). The methods described for gas
chromatographic analyses in toxic fats detected the
presence of octa-, hepta-, and hexachlorodibenzo-p-dioxins,
but the presence of lower chlorinated dioxins was masked
by interfering substances.

It has been suggested that the chick edema factor
might arise through contamination of fats with residual
chlorinated phenols which could give mixtures of dioxins
during production of fatty acids (Higginbotham et al.,
1968). Pyrolysis of a number of chlorophenols gave
mixtures of products identified as chlorinated dioxins.
Comparison of the retention times of dioxins in the
pyrolysates with those in a sample of toxic fat showed
that several of these were identical. Bioassay indicated
the toxicity of these pyrolysates to the chick embryo.
The authors concluded that, "When crude fats and tallows
are subjected to heating operations (hydrolysis, distillation in the production of commercial fatty acids residues
of commercial chlorophenols present in the fat might be
converted in part to chick edema factors." Since the
appearance of their publication, concentrated efforts
have been made to synthesize pure chlorinated dioxins,
to determine the dioxin content of commercial chlorinated
phenols and to examine the termal stability of some
chlorophenols.

Recently, the effects of thermal treatment on
individual chlorophenols were reported (Langer et al.,
1971, Stehl et al., 1971). The lower chlorinated phenols
heated in bulk or in solution did not produce measurable
amounts of dioxin. Pentachlorophenol gave some octachloro-
dibenzodioxin on prolonged heating above 200° C. The
sodium salts of the phenols, however, afforded reasonable
yields of dioxins. Sodium pentachlorophenate gave essentially
pure octachlorodibenzo-p-dioxin at 360°. Since this
compound is stable to 700° C, it could easily be recovered
from the thermal reactions. The sodium salts of the lower
chlorinated phenols also gave dioxins on pyrolysis, but the
formation of more highly condensed products was favored.

Chlorophenols and Dioxin Contaminants

The consequences of dioxin contamination are
important for the pesticide analytical chemist, not only
on account of the chick edema problem but because the use
of chlorophenols as pesticides or pesticide intermediates
has a direct bearing on product safety. Chlorophenols
are extensively used as bactericides or fungicides and
the use of pentachlorophenol for wood preservative
treatment is widespread. In addition to 2,4,5-T, we can
identify a number of pesticides based on chlorophenols in

which the structural arrangement of a chlorine atom
attached to a benzene ring in the ortho position to a
hydroxy group would permit dioxin formation. The
possibility of dioxin contamination during the production
of technical pesticides from these intermediates would
depend on the conditions necessary for synthesis. Therefore, an analytical survey of some technical compounds
was carried out to determine whether dioxins were present
(Table 2) (Woolson et al., 1971a). Fortunately, the
analytical survey showed that few of the compounds
examined were contaminated by dioxins. The lower
chlorinated phenols, such as 2,4-dichlorophenol used for
the manufacture of 2,4-dichlorophenoxyacetic acid (2,4-D),
are synthesized by chlorination of phenol under relatively
mild thermal conditions. By contrast, the highly
chlorinated phenols may be subjected to higher temperatures
during chlorination and these were generally found to
contain quantitatively more dioxins. The problem of
2,4,5-T merits individual discussion, because the starting
material, 2,4,5-trichlorophenol, is not prepared by direct
chlorination of phenol.

       TCDD is probably the most toxic member of the group
of chlorinated dioxins and is thus a major cause for
concern. However, there are over 70 chlorinated dibenzo-p-
dioxins and many possess toxicity approaching that of TCDD.
The related chlorinated dibenzofurans are also toxic and
possess about 1/20th of the toxicity of the dioxin with a
corresponding pattern of chlorine substitution (Kimmig
and Schultz, 1957). This factor of 1/20th signifies that
the problem of dibenzofurans cannot be considered
negligible if we consider that although the LD_{50} for chicks
is 60-80µg for tetrachlorodibenzofuran compared with 3-4 µg
for TCDD, the former figure is still a very small quantity.
The analysis of commercial chlorinated phenols, particularly
the non-phenolic portion, has therefore received much
attention from analytical chemists during the past two
years.

       Stehl et al. (1971b) of the Dow Chemical Company
have discussed this problem with regard to the problems
encountered in the analysis of 2,4,5-T and pentachlorophenol. The complex mixture of physically and chemically
similar compounds present in the non-phenolic fraction of
chlorinated phenols presents many analytical problems.
The non-phenolic portion represents only a small fraction

Table 2. Number and content of polychlorodibenzo-p-dioxins in selected pesticides (Woolson et al., 1971)

Pesticide	chlorodibenzo-p-dioxin (ppm)											No. of samples contaminated	Total No. of samples tested
	tetra-		hexa-			hepta-			octa-				
	<10	<100	<10	<100	<1000	<10	<100	<1000	<10	<100	<1000		
						Number found							
2,4,5-T	7‡	13	3	1	0	ND*	--	--	ND	--	--	23	42
silvex	1	0	ND	--	--	ND	--	--	ND	--	--	1	7
2,4-D (-DB,-DP)	ND	--	1	0	0	ND	--	--	ND	--	--	1	28
dicamba	ND	--	ND	--	--	ND	--	--	ND	--	--	0	8
chlorophenol													
tri-	ND	--	4	0	0	1	0	0	2	0	0	4	6
tetra-	ND	--	1	1	0	1	2	0	1	2	0	3	3
penta-	ND	--	0	7	0	0	4	6	0	4	6	10	11
others	ND	--	1	0	1	3	0	1	3	1	0	7	24

‡Any sample may contain 1 or more different dioxins.

*ND = <0.5 ppm of any one chlorodioxin.

of the total material present and its separation from
other compounds is difficult. Some of the sterically
hindered chlorinated phenoxyphenols do not appear to
dissolve in aqueous bases on extraction (Plimmer and
Klingebiel, 1971a). These compounds together with diphenyl
ethers, dibenzofurans and dioxins are present in the
mixture and their many possible patterns of chlorine
substitution add to the problem. Chlorine substitution
in compounds obtained during manufacture of 2,4,5-
trichlorophenol may also be accompanied by methoxy
substitution. Sample clean-up followed by single column
gas chromatography or multiple-column gas chromatography
have all been used for analysis sometimes in conjunction
with liquid chromatography. The combination of these two
techniques, if high pressure liquid systems can be used,
may provide a reasonably rapid and sensitive system for
handling small samples and is currently under study by the
Dow group.

To avoid the problems of sample clean-up, an extremely
specific method of detection might be employed. The
electron capture-gas detector responds to chlorinated
compounds present in the sample: the mass spectrometer
provides a more specific method of detection since the
fragmentation patterns are usually satisfactory guides to
the composition of the unknown. Separation by gas chromatography of the groups of chlorinated dioxins carrying
the same numbers of chlorine atoms can be easily achieved.
The examination of the column effluent by mass spectrometry
resolves the major problems of identity. Mass spectrometry
combined with gas chromatography is a powerful technique
and it is particularly useful for chlorinated compounds
since the number of chlorine atoms can readily be derived
from a measurement of the relative abundance of peaks
containing the isotopes ^{35}Cl and ^{37}Cl. By suitable
instrumental modification, the sensitivity of detection of
some individual dioxins can be increased so that the method
is capable of detecting several picograms of TCDD in a
sample.

The method is adequate for separation and detection
of TCDD and related compounds and surface coated open
tubular capillary (SCOT) columns may be used to improve
separation of closely related compounds. The fragmentation
pattern of TCDD is shown in Scheme 1 (Plimmer and Klingebiel,
1970). The long retention time of higher chlorinated dioxins

Scheme 1

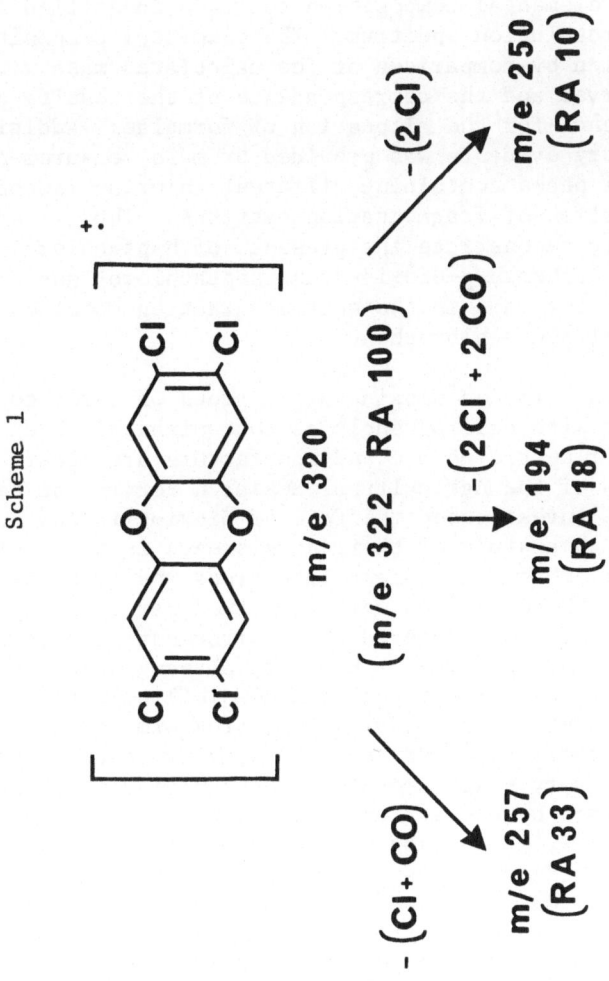

m/e 320

(m/e 322 RA 100)

-(Cl+CO) → m/e 257 (RA 33)

-(2Cl + 2CO) → m/e 194 (RA 18)

-(2Cl) → m/e 250 (RA 10)

and loss at the chromatograph-spectrometer interface have
limited the utility of this technique for analysis of
higher chlorinated dioxins. However, Plimmer et al. (1971)
have used the technique of direct introduction through the
solid sample probe of the high resolution mass spectrometer
to examine the neutral fraction of pentachlorophenol. Ions
of unique elemental composition could be identified in
the high-resolution spectrum. The elemental composition
was verified by comparison of the calculated mass with
that observed and the correspondence of the results provided
good evidence for the allocation of formulae. Additional
confirmatory evidence was provided by mass measurement of
individual peaks containing different chlorine isotopes and
by examination of fragmentation patterns. Thus it was
possible to demonstrate the presence of heptachloro and
octachloro dibenzo-p-dioxins and heptachloro- and octa-
chlorodibenzofurans in the neutral fraction obtained from
a sample of pentachlorophenol.

The chlorinated dioxins are a group of toxic compounds
associated with chlorophenols as contaminants. The implications for pesticide use and manufacture are clear.
Scientists of the Agricultural Research Service at Beltsville have investigated the fate of dioxins in the environment through a study of their persistence in soil, microbial
metabolism, movement, uptake into crops and decomposition
by light. These studies were concerned particularly with
TCDD which may have entered the environment as a contaminant
in the herbicide, 2,4,5-T. However, although the use of
pesticides can be regulated, the possibility that dioxins
could be formed in the environment from simple intermediates
cannot be overlooked and there is some evidence (Crosby et al.,
1971) that very small amounts of octachlorodibenzo-p-dioxin
may be formed by the action of light on pentachlorophenol.
By contrast 2,4-dichlorophenol does not give a dioxin by
irradiation (Plimmer and Klingebiel, 1971). Nor should the
possibility that preformed dioxins or furans may enter the
environment through other routes be forgotten. Vos et al.
(1970) showed that some polychlorinated biphenyls of
European manufacture and assayed on chromatography gave
fractions which were highly toxic to the chick embryo.
These fractions contained tetra and pentachlorodibenzofuran
which were identified by mass spectrometry.

It is important that we should be aware of the
properties of all chemicals released by man into the

environment regardless of whether they are pesticides.
At the USDA laboratories in Beltsville, we have also been
concerned with problems of analysis of dioxin in technical-
grade pesticides (Woolson et al., 1972) and also, in collaboration with the Department of the Interior, and the U. S. Air
Force with the analysis of dioxins in environmental samples
(Woolson et al., 1971b). No dioxins could be detected in
2,4,5-T treated soil samples or in tissue samples of bald
eagles taken from 12 states. In the case of pesticides,
the analytical chemist has a particular responsibility
to ensure that evidence for the identity of residues is
soundly based - preferably confirmation of identity by
several criteria should be attempted. These criteria may
frequently be a legal requirement to establish that a
pesticide is, in fact, present in a sample and, to take
broader view, the inclusion of misleading or incomplete data
in the pesticide literature will only provide future grounds
for fruitless controversy or entail useless diversion of
scientific effort.

Bibliography

1. Anonymous (1970). Effects of 2,4,5-T on man and the
    environment. Hearings before the Subcommittee on Energy,
    Natural Resources and the Environment of U.S. Senate
    Committee on Commerce, April 7 and 15th, 1970. Serial
    91-60 - U.S. Govt. Printing Office, Washington, D. C.

2. Anonymous (1971a). Report on 2,4,5-T: A report of the
    Panel on Herbicides of the President's Science Advisory
    Committee, Office of Science and Technology, March, 1971,
    Washington, D. C.

3. Anonymous (1971b). Restricting the Use of 2,4,5-T:
    Costs to domestic users. Agricultural Economic Report,
    No. 199 ERS and ARS, U. S. Dept. of Agriculture, March,
    1971, Washington, D. C.

4. Barlow, F. (1966). Nature 212 505 (1966).

5. Bitman, J. L., Cecil, H. C., and Fries, G. F. (1971).
    Science 174 64 (1971).

6. Cantrell, J. S., Webb, N. C. and Mabis, A. J. (1967).
    Paper presented at American Crystallographic Association
    Meeting, Atlanta, Georgia, Jan. 25-28, 1967. Chem.
    Eng. News 45 (5), 10 (1967).

7. Cantrell, J. S., Webb, N. C. and Mabis. A. J. (1969).
    Aeta Crystallogr. Sect. B. 1969, 150-6.

8. Crosby, D. G., Wong, A. S., Plimmer, J. R., and Woolson,
    E. A. (1971). Science 173 748 (1971).

9. Ecobichon, D. J. and Saschenbrecker, P. W. (1968).
    Can. J. Physiol. Pharmacol. 46, 785 (1968).

10. Federal Register (1961) No. 26 - 121:224, 121:1070.

11. French, M. C. and Jefferies. D.C. Science 165 914
    (1969).

12. Friedman, L., Firestone, D., Horowitz, W., Banes, D.,
    Anstead, M., and Shue, G. J. Assoc. Offic. Agric.
    Chem. 42 129 (1959).

13. Gardner, A. M., Damico, J. N., Hanse, E. A., Lustig,
    E. and Storrherr, R. W. (1969). J. Agr. Food Chem.
    17 1181-1185 (1969).

14. Graham, R. E. and Kenner, C. T. (1969). J. Agr.
    Food Chem. 17 259-263 (1969).

15. Haller, H. L., Bartlett, P. D., Drake, N. L., Newman,
    M. S., Cristol, S. J., Eaker, C. M., Hayes, R. A.,
    Kilmer, C. W., Magerlein, B., Mueller, C. P., Schneider,
    A., and Wheatley, W. (1945). J. Am. Chem. Soc. 67
    1591-1603 (1945).

16. Herxheimer, K. (1899). Munch. Med. Wschr. 278 (1899).

17. Kearney, P. C., Isensee, A. R., Helling, C. S., Woolson,
    E. A., and Plimmer, J. R. 162nd ACS National Meeting,
    Washington, D. C., September 13-17, 1971, Div. of
    Pesticide Chem., Abs. No. 90.

18. Kimmig, J. and Schulz, K. H. (1957). Dermatologica
    115, 540 (1957).

19. Klein, A. K. et al. J. Assoc. Offic. Anal. Chem. 47 1129 (1964).

20. Klein, A. K., Lang, E. P., Datta, R. P., and Mendel, J. L. (1965). J. Am. Chem. Soc. 87 2520 (1965).

21. Langer, H. G., Brady, T. P., Dalton, L. A., Shannon, T. W., and Briggs, P. R. (1971). Thermal Chemistry of Chlorinated Phenols. 162nd ACS National Meeting, Washington, D. C., Sept. 13-17, 1971. Div. of Pesticide Chem., Abs. No. 83.

22. Neal, P. (1967). J. Assoc. Offic. Anal. Chem. 50 1338 (1967).

23. Plimmer, J. R., and Klingebiel, U. I. (1970). Unpublished observations.

24. Plimmer, J. R., and Klingebiel, U. I. (1971a). Unpublished observations.

25. Plimmer, J. R., and Klingebiel, U. I. (1971b). Science 174 407 (1971).

26. Plimmer, J. R., Ruth, J. M., and Woolson, E. A. Unpublished observations.

27. Schechter, M. S. (1968). Pesticide Monitoring J. 2 1 (1968).

28. Segal, H., in G. Zweig, ed., "Pesticides Plant Growth Regulators and Food Additives", Vol. 5, Academic Press, Inc., New York, 1967, pp. 321-334.

29. Stehl, R. H., Papenfuss, R. R., Bredeweg, R. A., and Roberts, R. W. (1971a). 162nd ACS National Meeting, Washington, D. C., Sept. 13-17, 1971, Div. of Pesticide Chem., Abs. No. 92.

30. Stehl, R., Wilke, E., Papenfuss, R., Matalon, R. (1971b). 162nd ACS National Meeting, Washington, D. C. Sept. 13-17, 1971, Div. of Pesticide Chem., Abs. No. 81.

31. Tishler, M. (1960). Private communication quoted in Yartzoff et al., (1961).

32. Vos, J. G., Koeman, J. H., Van Der Maas, H. L., Ten
     Noever De Brauw. M. C., De Vos, R. H. (1970). Fd.
     Cosmet. Toxicol., 8 625-633 (1970).

33. Woolson, E. A., Reichel, W. L., and Young, A. L. (1971b).
     162nd ACS National Meeting, Washington, D. C. Sept. 13-
     17, 1971, Div. of Pesticide Chem., Abs. No. 91.

34. Woolson, E. A., and Thomas, R. F. (1971). Proceedings
     of the 2nd International Pesticide Congress IUPAC,
     Tel Aviv, Israel, Feb. 1971 (In Press).

35. Woolson, E. A., Thomas, R. F., and Ensor, P. D. J.
     (1972). J. Agric. Food Chem. 20 351-354.

36. Wooton, J. C., and Courchene, W. L. (1964). J. Agr.
     Food Chem. 12 94-98 (1964).

37. Yartzoff, A., Firestone, D., Banes, D., Horwitz, W.,
     Friedman, L. and Nesheim, S. (1961). J. Am. Oil
     Chem. 38 60-62 (1961).

GAS CHROMATOGRAPHIC ANALYSIS OF PESTICIDE RESIDUES CONTAINING
PHOSPHORUS AND/OR SULFUR WITH FLAME PHOTOMETRIC DETECTION
AND SOME ANCILLARY TECHNIQUES FOR VERIFYING THEIR IDENTITIES

M. C. Bowman

U. S. Department of Agriculture, Agricultural
Research Service, Tifton, Georgia   31794

In our efforts to diminish insect depredation of food
and fiber, insecticides continue to be our primary weapon.
The persistent chlorinated hydrocarbons have now been
largely replaced by organophosphorus and carbamate insecticides because they are bio-degradable and do not tend to
accumulate in animal tissues and to generally pollute the
environment.

We have been using the Melpar flame photometric detector
(Brody and Chaney 1966)(Tracor, Inc., Austin, Texas) for
gas chromatographic determination of residues of phosphorus
(P)- and/or sulfur (S)-containing pesticides and insect
chemosterilants for about 5 years. During this time, more
than 100 compounds have been analyzed in about 50 different
substrates. Because of the detector's highly specific and
sensitive response to P and S, little or no cleanup was
required. Subnanogram quantities of P compounds and a few
nanograms of S compounds are detectable (twice noise).
These qualities and the fact that it can be easily temperature-
programmed indicated that the detector was especially
suitable for the analysis of samples containing pesticides
with a wide range of volatilities and thus applicable to
environmental samples having an unknown history of treatment.

Based on the results of our studies during the past
several years, we have proposed a method for the multicomponent analysis of pesticide residues containing P and
S (Bowman et al., 1971). In this short communication I can

175

only present the basic principles of the method and acquaint
you with what can be done. It will be necessary to consult
the reference for the details of this and the other procedures
to be described. The method may be divided into several
parts: extraction, cleanup, GC analysis, and confirmation
of identity.

The extraction procedure is a key step in an analysis
for pesticide residues; yet, it is often the one given
minimal consideration by the analyst. Surface stripping
and blending techniques which may provide complete recoveries
of pesticides from fortified samples often yield low recoveries
when applied to field-weathered crops (Bowman et al., 1968b).
For example, with a benzene blend, recovery of Ciba C-9491
O-(2,5-dichloro-4-iodophenyl)O,O-dimethyl phosphorothioate
and two metabolites from field-weathered Coastal bermudagrass
was only about half that obtained by Soxhlet extraction
with chloroform - 10% methanol (Bowman and Young, 1969).
The latter method was therefore adopted for routine use in
our laboratory. The extraction is performed under an
atmosphere of N_2 for the length of time (usually about 4 hours)
required to recover all detectable residues extractable by
the method. Liquid samples or those not amenable to the
Soxhlet procedure are exhaustively extracted by blending with
acetone (usually several times).

In the cleanup, the raw dry extract is partitioned
between the hexane-acetonitrile pair to transfer the fat
and other non-polar constituents into the hexane layer;
the acetonitrile is then injected directly into the gas
chromatograph (GC). Sometimes the pesticide and metabolites
do not partition sufficiently into the acetonitrile layer;
in these instances the hexane layer is extracted several
times with fresh acetonitrile or corrections based on
partition values (p-values) are applied (Bowman and Beroza ,
1965, Beroza et al., 1969).

The analysis is performed with a GC equipped with the
flame photometric detector modified for high temperature
operation (Bowman et al., 1968a, Dale and Hughes, 1968)
and programmed from 150 to 300°C at 10°C/min, then held at
300°C until all pesticide peaks emerge. Sensitivities are
generally 0.01-0.05 ppm for P compounds and 0.02-0.10 ppm
for S compounds. Temperature programming improves the
sensitivity of late-eluting pesticides because their peaks
are tall and narrow compared to those obtained isothermally.

Retention times of 146 P and/or S pesticides, metabolites, and insect chemosterilants relative to parathion on five thermally stable columns (OV-101, OV-17, OV-210, OV-225, and Dexsil 300) have been reported (Bowman and Beroza, 1970, 1971a; Bowman et al., 1971). This compilation aids the analyst in tentatively identifying GC peaks suspected of being pesticides or in selecting the proper column for resolving pesticide peaks. As an additional aid to the analyst, temperature-programmed chromatograms of 39 foods on four of the columns in both the P and S modes of detection were reported (Bowman et al., 1971). (Although chromatograms of the foods on Dexsil 300 have not been reported, they are expected to be very similar to those from OV-101). By locating the retention time of a pesticide on the chromatogram of a food, the analyst may easily ascertain whether interference can be expected, and he can decide whether a particular analysis may be accomplished without additional cleanup or whether another liquid phase should be selected.

Typical chromatograms of pesticide standards are presented in Figure 1 and chromatograms of food extracts with the OV-101 packing and the P mode are given in Figure 2. Residues of P-containing pesticides may be determined with little or no interference in more than 80% of the foods. Chromatograms of the foods on the same packing using the S detector are presented in Figure 3. These are less satisfactory than with the P detector primarily because larger injections of extracts are required to offset the lesser sensitivity of the detector in the S mode. The resulting buildup of sample extractives on the column tends to shorten column life and cause interference, such as baseline elevation, to appear in subsequent temperature programmed analyses. This problem was solved, at least partly, by using a packing of Dexsil 300 which has a very high thermal stability (Bowman and Beroza, 1971a). Contaminated columns were readily restored to service by replacing the first 12 cm. of the packing and purging the column at 400°C for 1 hour. The silicone rubber O-rings generally used for sealing glass columns were not satisfactory for use at 400°C; they had to be replaced by gaskets prepared from asbestos impregnated with Dexsil 300 (Beroza and Bowman, 1971).

Methods for confirming the identities of GC peaks are required to bolster the integrity of residue results, particularly in regulatory laboratories or when the history of treatment is not known. Since the analytical sample generally consists of nanogram quantities of the pesticide,

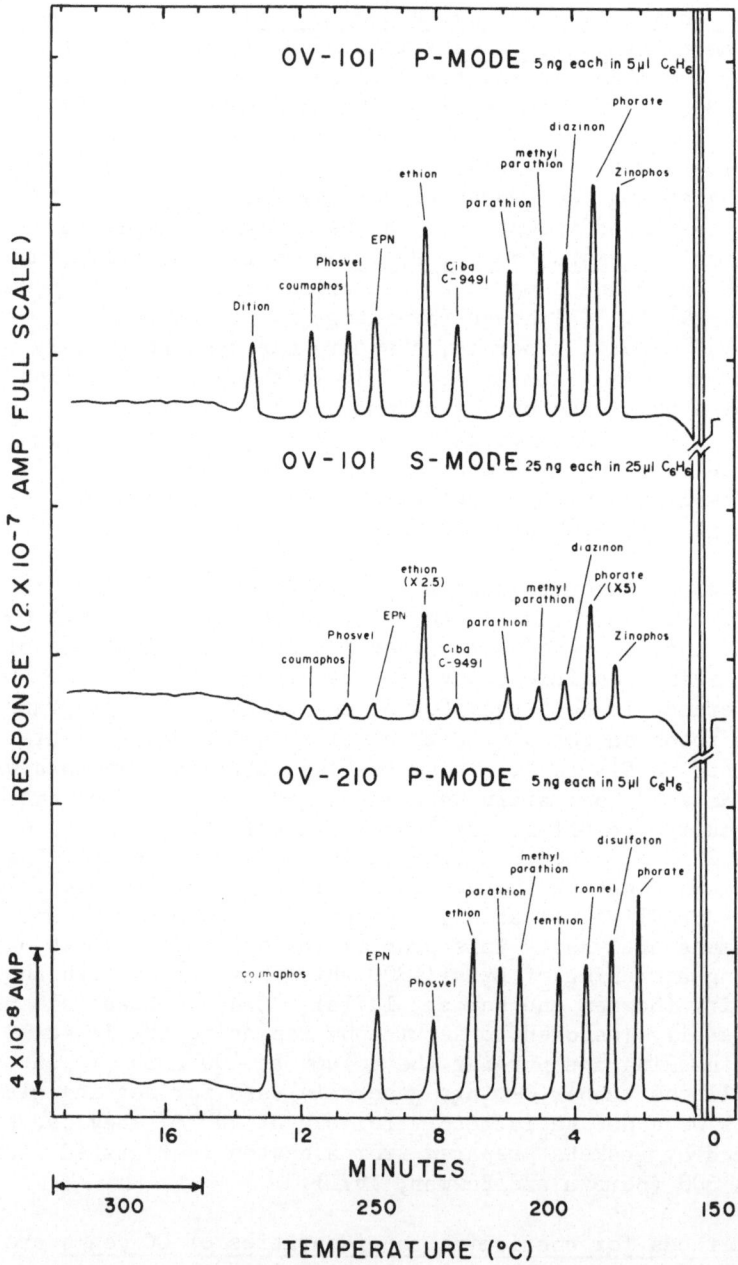

Figure 1.   Chromatograms of pesticide standards in
multi-component residue method
(Bowman et al., 1971).

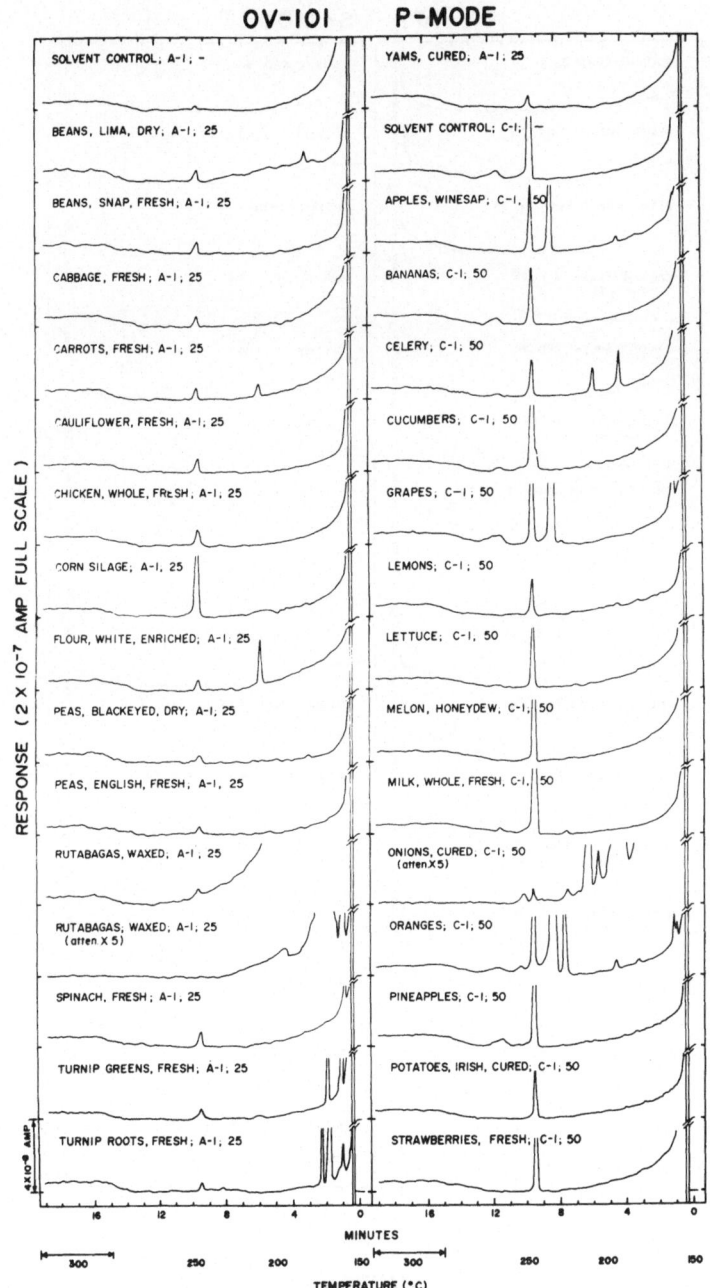

Figure 2. Chromatograms of extracts of foods in multi-
           component residue method using OV-101 packing and
           P detector (Bowman et al., 1971).

OV-101   S-MODE

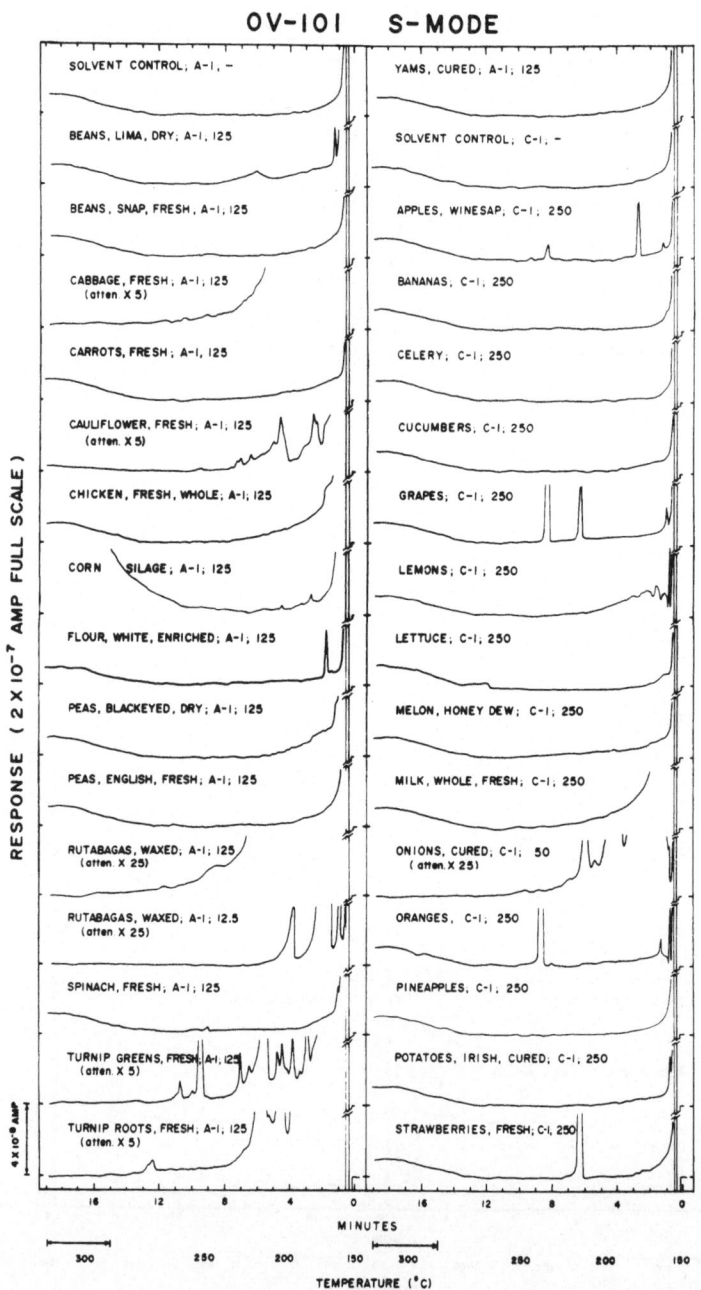

Figure 3.   Chromatograms of extracts of foods in multicomponent residue method using OV-101 packing
and S detector (Bowman et al., 1971).

usually in combination with large amounts of extraneous
material, techniques such as spectrophotometry, mass
spectrometry, and NMR are not readily applicable for confirmation of identity without a rigorous cleanup; furthermore the
quantity of pesticide is most often insufficient for an
analysis.

One of the most common methods currently employed is a
simple comparison of retention time of the unknown peak
versus peaks of standard pesticides on several different
types of column packings such as those already mentioned.
An identical retention time of a standard and unknown peak
on two or more GC columns constitutes good supplementary
evidence that the compounds are the same.

The dual channel flame photometric detector (Bowman and
Beroza, 1968) which simultaneously monitors P (526 nm
filter) and S (394 nm filter) compounds in the GC effluent
also provides good ancillary information for confirming the
identity of a GC peak. In addition to qualitative evidence
concerning the presence (or absence) of P and/or S in the
compound generating the peak, the atomic ratio of P to S
in the molecule may also be estimated.

A typical temperature-programmed analysis of 9 pesticides
using the dual detector is presented in Figure 4. Gardona®,
which contains P but no S, is detected on the P channel but
not on the S channel.® A similar analysis is illustrated in
Figure 5. Sulphenone®, which contains S but no P, responds
only on the S channel. An investigation of the relative
response of the two channels for compounds containing both P
and S revealed that the atomic ratio of these elements may
be determined by dividing the P response by the square root
of the S response. Table 1 gives such response ratios for
21 pesticides at three levels of concentration (100, 50, and
25 ng). The response ratios remain fairly constant for any
given compound, and these ratios are related to the PS contents
of the molecules. For example, the ratios at the three
concentration levels for Zinophos®, a PS compound, range
between 5.5 and 6.0 as do the ratios of all other pesticides
containing one P and one S. Ratios of the PS_2 pesticides
are about 3 (ca. half that of the PS compounds), and those
with a PS_3 content are about 2 (ca. one-third that of the
PS compounds). Of course, by reverse application of this
process, the analyst can deduce the atomic ratio of P to S
in an unknown compound. The capability of obtaining results

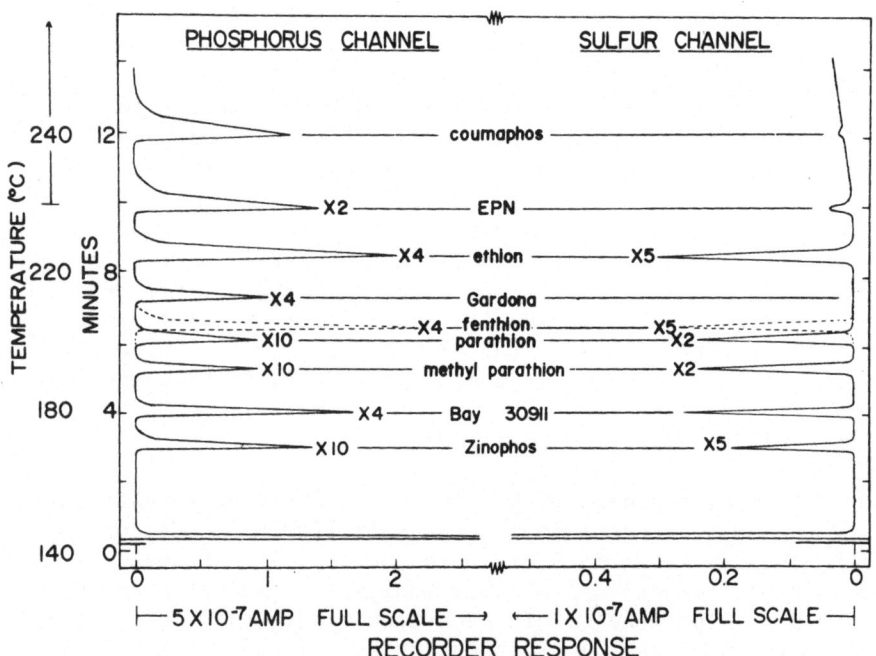

Figure 4. Dual-channel recording of P and S response to
          50-ng amounts of 9 pesticides in 5 μl of benzene
          in temperature-programmed GC (Bowman and Beroza, 1968

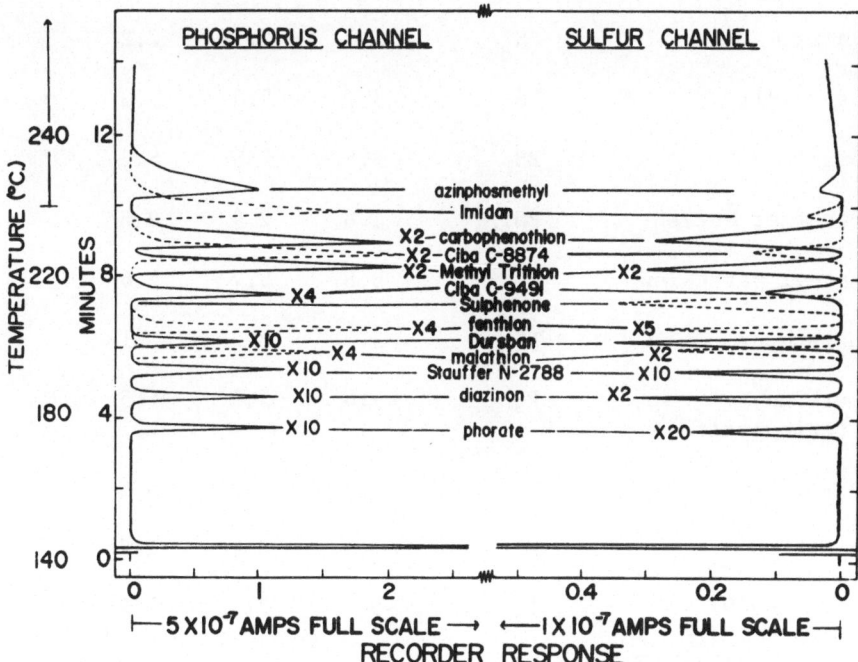

Figure 5.   Dual-channel recording of P and S response to 13
             pesticides analyzed as described in Figure 4
             (Bowman and Beroza, 1968).

Table 1. Response ratios ($R_p / \sqrt{R_s}$) pf compounds containing
phosphorus and sulfur

	PS content	Response X 10^3 at concentration (ng) indicated		
		100	50	25
Zinophos®	PS	5.5	5.8	6.0
Phorate	PS_3	2.0	2.1	2.2
Bay 30911	PS	5.7	5.5	5.3
Diazinon	PS	5.5	5.6	5.3
Stauffer N-2788	PS_2	2.8	2.9	2.8
Methyl parathion	PS	5.8	5.7	5.6
Malathion	PS_2	3.0	3.2	3.1
Dursban® (chlorpyrifos)	PS	5.6	5.6	5.5
Parathion	PS	5.6	5.5	5.5
Fenthion	PS_2	3.0	3.0	2.9
Ciba C-9491	PS	5.3	5.2	5.2
Methyl Trithion®	PS_3	1.9	2.0	1.9
Ethion	P_2S_4	2.8	2.9	2.8
Ciba C-8874	PS	5.4	5.2	5.3
Carbophenothion	PS_3	1.9	2.0	1.8
Imidan®	PS_2	3.3	3.2	3.3
EPN	PS	6.0	5.8	-
Azinphosmethyl	PS_2	3.2	3.3	-
Coumaphos	PS	5.9	-	-

of this type can be most important in confirming the identity
of GC peaks containing P and S.

The p-value is defined as the fraction of solute
partitioning into the non-polar phase of two immiscible
phases of equal volume. The p-value is easily and rapidly
determined from a single distribution, is especially useful
for confirming the identity of pesticide residues at levels
amenable to quantitative analysis by electron-capture and
flame photometric gas chromatography. Each pesticide has a
characteristic p-value with a given solvent pair, and it is
not necessary to determine the amount of substance in the
analysis, only the relative amounts present in the original
and the extracted solution. This is especially convenient
when dealing with an unknown compound for which the response
factor is not known. p-Values for 31 P- and/or S-containing
compounds and metabolites in hexane-acetonitrile are presented
in Table 2; additional values in four solvent systems for
several pesticides (and metabolites) containing a thioether
linkage are presented in Table 3. The p-values are sufficiently reproducible and different from one another to allow
their use in confirming the identity of pesticides to be
meaningful.

Popularity of the p-value procedure as a means of
confirming the identity of GC peaks at the nanogram level is
probably exceeded only by the technique of comparing retention
times on different columns. The integrity of a p-value
confirmation is greatly enhanced by using several different
solvent systems; more than twenty such binary systems have
been used (Beroza and Bowman, 1965, Bowman and Beroza, 1966).

Another excellent technique for confirming identity of
many of the P and/or S pesticides and metabolites concerns
their behavior in liquid chromatography. Many of the
compounds contain either a P=S or P=O (O-analog) moiety;
others also contain a thioether linkage. For example,
(Figure 6) the phosphorothioate insecticide, fenthion (I)
can be oxidized to the O-analog (IV). In addition, the
sulfide in the para position of the aromatic ring of both
compounds can be oxidized to sulfoxides (II, V) or sulfones
(III, VI). Thus, the use of fenthion can result in residues
of any or all of the six compounds. These compounds differ
widely in polarity (compare their p-values in Table 3);
O-analogs are more polar than the corresponding thiophosphates;
sulfoxides are more polar than sulfones, which are more polar

Table 2. p-Values of 31 P and/or S pesticides and metabolites
in hexane-acetonitrile at 25°C.

Azinphosmethyl	.008	Gardona ®	.051
Bidrin ® (dicrotophos)	.008	Imidan ®	.007
Carbophenothion	.21	Malathion	.042
Chlorophoxim	.047	Methyl parathion	.022
Chlorophoxim O-analog	.011	Methyl Trithion ®	.075
Ciba C-9491	.26	Nemacur ®	.038
Coumaphos	.006	Parathion	.044
Diazinon	.28	Phorate	.21
Disulfoton	.20	Phosvel ® (leptophos)	.35
Dursban ® (chlorpyrifos)	.28	Phosvel O-analog	.10
EPN	.038	Phosvel phenol	.10
Ethion	.079	Phoxim	.056
Fenitrothion	.035	Phoxim O-analog	.012
Fenitrothion O-analog	.007	Ruelene ® (crufomate)	.031
Fenitrothion cresol	.012	Shell Compound 4072	.058
		Zinophos ®	.058

Table 3. p-Values of pesticides containing a thioether
linkage (and their metabolites) in four
solvent systems.

Compound	Hexane-water	Hexane-20% MeCN (80% water)	Hexane-40% MeCN (60% water)	Benzene water
Fenthion	1.00	0.98	0.92	–
Fenthion sulfoxide	.50	.18	.03	–
Fenthion sulfone	.94	.61	.12	–
Fenthion O-analog	.92	.65	.18	–
Fenthion O-analog sulfoxide	.00	.00	.00	0.35
Fenthion O-analog sulfone	.01	.00	.00	1.00
Disulfoton	1.00	1.00	1.00	–
Disulfoton sulfoxide	.50	.23	.06	–
Disulfoton sulfone	.82	.52	.13	–
Disulfoton O-analog	.83	.75	.39	–
Disulfoton O-analog sulfoxide	.00	.00	–	–
Disulfoton O-analog sulfone	.01	.00	.00	.18
Oxydemetonmethyl	.00	.00	–	.08
Oxydemetonmethyl sulfone	.00	.00	–	.10
Phorate	1.00	1.00	1.00	–
Phorate sulfoxide	.54	.35	.10	–
Phorate sulfone	.98	.79	.30	–
Phorate O-analog	.89	.73	.36	–
Phorate O-analog sulfoxide	.00	.00	.00	–
Phorate O-analog sulfone	.01	.00	.00	.21
Mesurol®	1.00	.91	.43	1.00
Mesurol sulfoxide	.00	.00	.00	.48
Mesurol sulfone	.00	.00	.00	.96
Mesurol phenol	1.00	.86	.40	1.00
Mesurol phenol sulfoxide	.00	.00	.00	.06
Mesurol phenol sulfone	.00	.00	.00	.31
Dasanit®	.75	.34	.06	1.00
Dasanit sulfone	.98	.72	.18	1.00
Dasanit O-analog	.00	.00	.00	.62
Dasanit O-analog sulfone	.02	.02	.01	.96

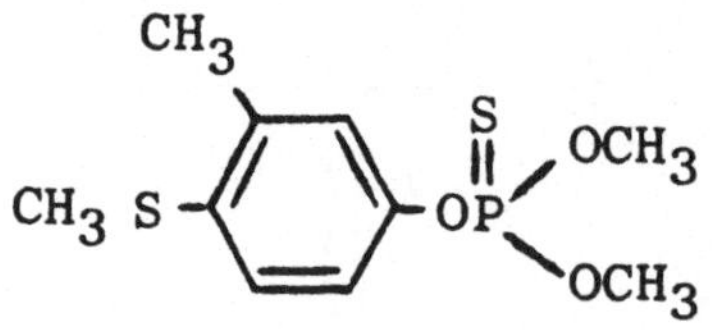

Figure 6.   Fenthion (I) and five of its metabolites.

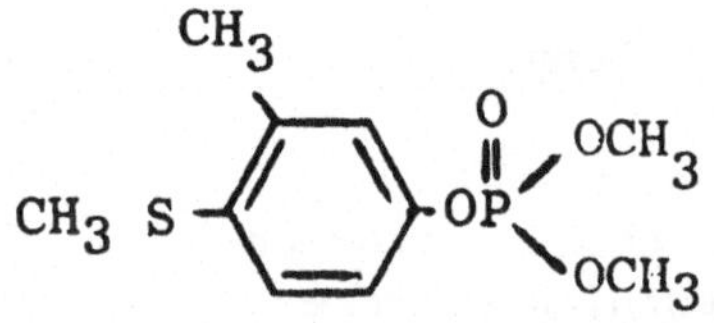

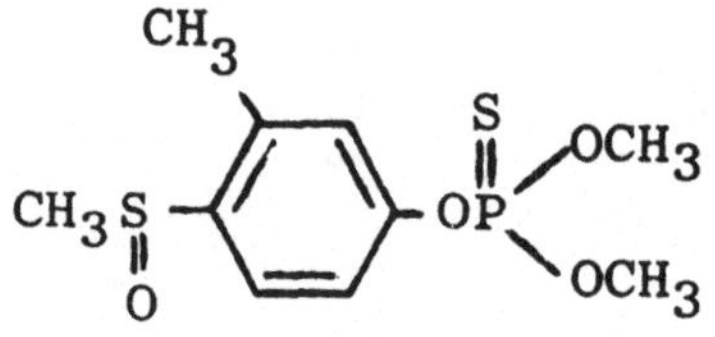

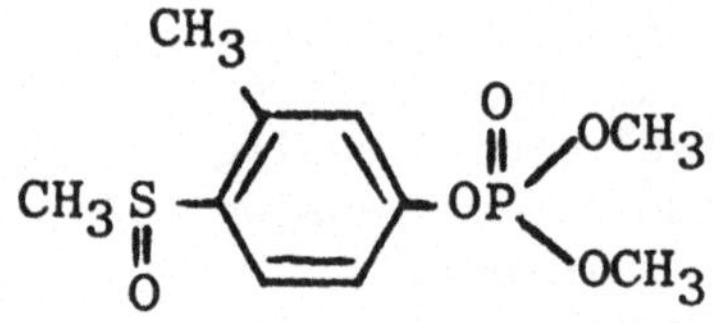

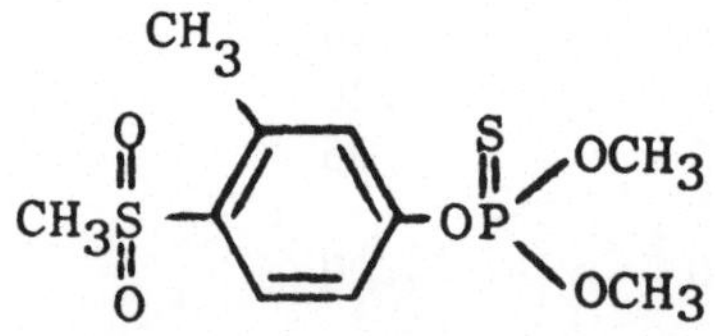

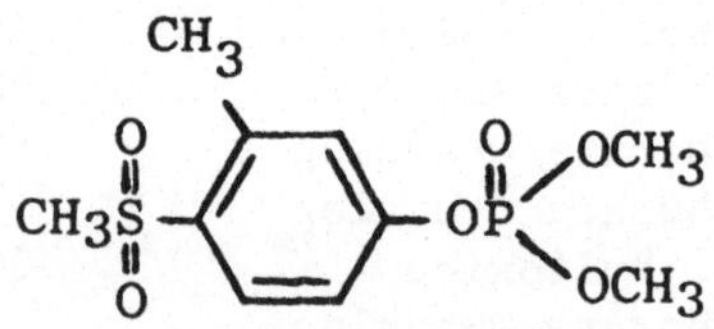

than sulfides. Behavior of 100 µg quantities of each of the six compounds on a 4 g silica gel column (contains 3.5% water) is illustrated in Figure 7. All six compounds separate as the column is developed with increasingly polar mixtures of benzene-acetone. Similar separations have been achieved for phorate, disulfoton, Dasanit®, Nemacur®, and Mesurol®, and their metabolites and many P=S, P=O pesticide mixtures.

Accordingly, the elution characteristics on silica gel for compounds with these functional groups are suggested as a means of confirming identity.

If a tentative identification indicates a residue exists at less than its highest oxidation state (e.g., fenthion or metabolites II, III, IV, or V, see Figure 6), then conversion of the residue to the O-analog sulfone by adding m-chloroperbenzoic acid (Bowman and Beroza, 1969) can further substantiate the identification.

Finally, we have to recognize that the problems associated with a general method for multicomponent residue analysis are formidable, and that many problems remain unsolved. We hope that pesticide residue chemists will find the methods proposed useful in their analyses of environmental samples and that they will devise improvements of the procedures being advanced.

SUMMARY

Gas chromatography with flame photometric detection is a highly specific and sensitive means of analyzing a wide variety of crops and tissues for pesticides, chemosterilants, and other compounds containing phosphorus (P) and/or sulfur (S). Through the use of temperature programming, multicomponent analyses for these residues are made with good sensitivity and with little or no cleanup. A compendium of relative retention times of 146 pesticides on 5 thermally stable packings and chromatograms of 39 foods was prepared to facilitate the identification and determination of pesticides in a wide range of samples and in those of unknown history. The dual detector simultaneously monitors the GC effluent for compounds containing P (526 nm filter) and S (394 nm filter); from the response of its two channels the P/S ratio of a compound is readily ascertained. p-Values and behavioral characteristics of these compounds

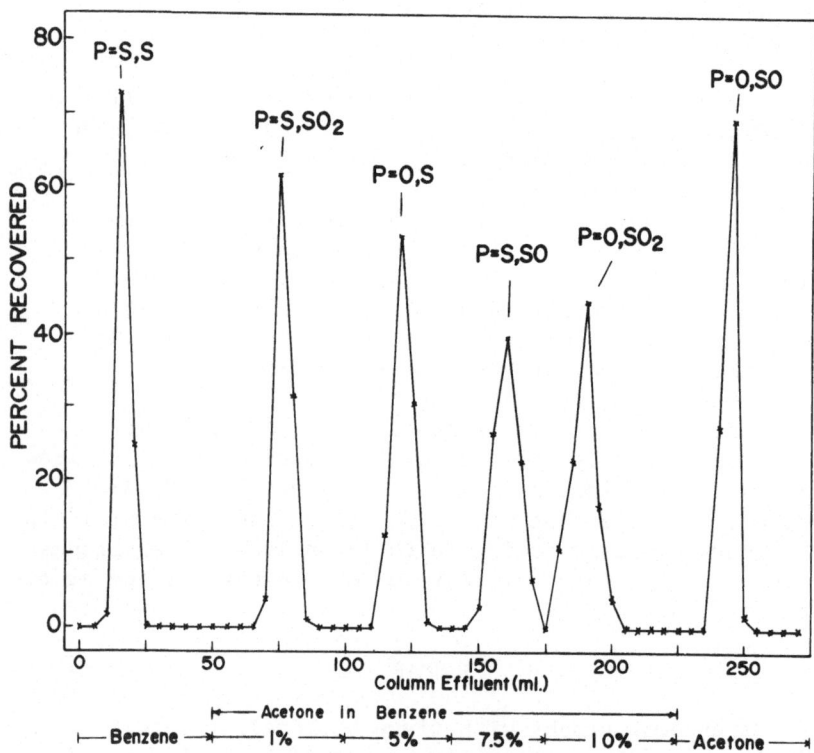

Figure 7.   Separation of fenthion and five of its metabo-
            lites by liquid chromatography on silica gel.
            (Bowman and Beroza, 1968a).

in liquid adsorption chromatography as ancillary means for
confirming identities of pesticides are discussed.

REFERENCES

Beroza, M., and Bowman, M. C. J. Ass. Offic. Anal. Chem. 48,
    943-52 (1965).

Beroza, M., Inscoe, M. N., and Bowman, M. C. Res. Rev. 30,
    1-61 (1969).

Beroza, M., and Bowman, M. C. Anal. Chem. 43, 808 (1971).

Bowman, M. C., and Beroza, M. J. Ass. Offic. Anal. Chem.
    48, 943-52 (1965).

Bowman, M. C., and Beroza, M. Anal. Chem. 38, 1544-49 (1966).

Bowman, M. C., and Beroza, M. Anal. Chem. 40, 1448-52 (1968).

Bowman, M. C., and Beroza, M. J. Agr. Food Chem. 16, 399-
    402 (1968a).

Bowman, M. C., Ford, H. R., Lofgren, C. S., and Weidhaas, D. E.
    J. Econ. Entomol. 61, 1586-89 (1968).

Bowman, M. C., Leuck, D. B., and Beroza, M. J. Agr. Food
    Chem. 16, 796-802 (1968a).

Bowman, M. C., and Beroza, M. J. Ass. Offic. Anal. Chem. 52,
    1231-37 (1969).

Bowman, M. C., and Young, J. R. J. Econ. Entomol. 62,
    1468-72 (1969).

Bowman, M. C., and Beroza, M. J. Ass. Offic. Anal. Chem. 53
    499-508 (1970).

Bowman, M. C., Beroza, M., and Hill, K. R. J. Ass. Offic.
    Anal. Chem. 54, 346-58 (1971).

Bowman, M. C., and Beroza, M. J. Ass. Offic. Anal. Chem.
    54, 1086 (1971a).

Brody, S. S., and Chaney, J. E.   <u>J</u>. <u>Gas</u> <u>Chromatogr</u>. <u>4</u>,
    42-6 (1966).

Dale, W. E., and Hughes, C. C.   <u>J</u>. <u>Gas</u> <u>Chromatogr</u>.   <u>6</u>, 603-4
    (1968).

The Determination of
Anions in Water

Edited by Jack L. Lambert

INTRODUCTORY REMARKS: SOME PARTIALLY SOLVED PROBLEMS IN
ANION ANALYSIS

Jack L. Lambert

Kansas State University

Manhattan, Kansas 66502

While instrumental methods of analysis occupy our
attention to an increasing degree these days, we find the
traditional wet chemical methods solidly entrenched in the
area of water analysis. A count of the methods listed in
the 12 and 13th editions of Standard Methods for the Examination of Water and Wastewater, excluding methods for non-
ionic organic compounds and radioactive species, shows the
following distribution

	Colorimetric	Titrimetric	Gravimetric	Instrumental
13th Ed.	66	20	6	28
12th Ed.	61	25	6	6

The methods used for anion analysis can be categorized as
follows

	Colorimetric	Titrimetric	Gravimetric	Instrumental
13th Ed.	29	9	3	4
12th Ed.	29	9	3	2

Instrumental methods are defined as those in which the sam-

ple receives little or no chemical pretreatment. The
single turbidimetric method is counted as a colorimetric
method.

Closer study of the anion methods shows that one anion
- vanadate - has been added and determined by a tentative
colorimetric method. A tentative colorimetric method for
selenium involving its reduction to an elemental sol and
a tentative polarographic method for nitrate have been
dropped. A tentative second modification of the methylene
blue method for sulfide has been added. A tentative colo-
imetric method for nitrate involving reduction by cadmium
amalgam preliminary to a diazotization procedure has been
added as an alternative to the tentative zinc reduction
procedure carried over from the 12th edition. Another
tentative colorimetric method added for nitrate is the
chromotropic acid method. Instrumental methods added
include the fluoride-selective electrode method and gas
chromatographic methods for phenols and volatile organic
acids.

Admittedly, the definition of instrumental methods
is an arbitrary one, as every analytical method involves
measurement by some instrument, if only the human senses.
Lest we forget, however, the senses of taste and smell
stand unchallenged in the field of flavor and odor determination, and the properly calibrated human is a wonderful
and expensive instrument. The rather crude human neuro-
muscular sense long ago gave way to the balance for the
determination of weight, but the human eyeball has only in
this century been supplanted by the spectrophotometer. We
have ranged far beyond the limited photon response of the
retina, and have developed sensors for processes we can
never directly sense.

Although I argue the case for more research on all
types of methods, we must recognize the special need for
more and better instrumental methods. We are currently
at the stage where we can incorporate analytical instruments into some computer-controlled feed-back processes
and thus eliminate the human element in quality control.

We will be working, therefore, eventually to abolish our
profession but it is a distinguishing characteristic of a
profession, in contrast to some trades, to work toward its
own demise if that is the logical course. Dentists supported fluoridation of water supplies even though it would
eventually mean less demand for their services. Some
might wish that the legal profession would take the same
altruistic attitude toward no-fault automobile insurance.
If we must debate whether analytical chemistry is, or is
not, a fading discipline, we must first establish some
ground rules. If we mean that there will be fewer human
analysts in the future in proportion to the total population, the answer almost certainly is yes. If we mean that
there will be less need for analyses, the answer is no.
As analytical chemists we may be fading away but we plan
to leave a legacy of instrumental methods as a memorial.
However, we will be around for some time, so condolences
are not yet in order.

In the current revival of anti-intellectualism, analytical chemistry has probably suffered less than any other
branch of the physical sciences. We can't be blamed for
dirtying up the environment and we're certainly needed if
it is to be cleaned up. The ecology push may be our Sputnik, and we may look back on this as the golden age of analytical chemistry. If we have a complaint, it is the fact
that non-scientists (including engineers) control the
spending of funds for environmental research and development. The point was made at the recent National Bureau of
Standards symposium on analytical chemistry in national
problem areas that some of the most influential advisory
panels on pollution research have no analytical represent-
atives.

To focus on the topic of this session, we must make
the point that the range of analytical techniques applicable to anion analysis is limited in comparison to those
available for metal cations. There was generally less
interest in anion analysis until the recent introduction
of ion-selective electrodes but that interest centers on
the tools and not the species. We can expect to see some

established methods supplanted by anion-selective electrodes but probably not as many as some enthusiasts anticipate. There will still be a need for some new and better
spectrophotometric, spectrofluorometric, and even titri-
metric methods so long as there are analysts around to do
them.

I feel no concern about the future of instrumental
methods of analysis. The research on new methods is in
competent and energetic hands. I am seriously concerned
about the future of research on wet chemical methods. The
training of all types of chemists has become very restricted in recent years. Except for organic chemists, who
themselves are showing strong tendencies to become physical
chemists, reaction chemistry has largely dropped out of the
curriculum. I blame this state of affairs on the post-
Sputnik science boom when research productivity in a narrow specialty came to be emphasized at the expense of broad
and thorough training even within the graduate student's
major area. There is faint encouragement in the fact that
some faculties are beginning to reconsider their goals and
policies in view of the decreased demand for their product.

What are some of the partially solved problems in
anion analysis? First on my list would be sulfate determination. If ever there was an inert species, it is sulfate ion. It can't be oxidized, it isn't easily reduced
– in fact, but for its precipitation reactions with barium
and benzidinium type cations, we would have virtually no
handles on the problem. The existing methods are adequate
for the concentration ranges normally encountered but not
economical in terms of the analyst's time. It is fortunate
we do not need precise methods for the sub-parts per million range. Still it is a challenge to the research
analyst.

If ever the profession could point to a job well done,
it would be the development of methods for fluoride ion.
For control of fluoridation, analysts were called on to
devise methods that could be used by water works personnel
to measure fluoride concentrations at 1.0 ± 0.1 ppm. The

success of these efforts has received all too little recognition. With few exceptions, these methods involve visual or spectrophotometric measurements in solutions undergoing change in hue. No completely successful method incorporating formation of a single color proportional to fluoride concentration has yet appeared. It will be interesting to see if the fluoride-selective electrode discourages further research on new spectrophotometric methods.

It seems to me that we have need for more and better simple methods of analysis. There are on the market a variety of test kits and indicator papers for chemists and interested non-chemists such as rural and suburban homeowners and hikers who want to monitor their own water supplies. There are few cumulative sampling techniques available to detect traces of pollutants that might find their way intermittently into surface and underground waters. Cumulative sampling devices have been used for air pollutants, and the idea might profitably be carried over into water analysis. The research involved could be very interesting, especially if new methods were to be developed rather than adaptations of existing methods.

In summary, there is life in the profession still. If we can convince the knowledgeable non-scientists that all the analytical problems connected with the environment are not yet solved, we can make significant contributions to the fight against pollution. We should seriously consider whether we are training our analytical chemists properly. Certainly those who do research on analytical methods, of all chemists, should have a broad background in all areas of chemistry.

SPECTROPHOTOMETRIC, SPECTROFLUOROMETRIC AND ATOMIC ABSORPTION SPECTROMETRIC METHODS FOR THE DETERMINATION OF ANIONS IN WATER

David F. Boltz

Department of Chemistry, Wayne State University, Detroit, Michigan 48202

INTRODUCTION

In surveying for many years the inorganic and organic analytical chemistry sections of Chemical Abstracts, it has been quite apparent that the attention given to the determination of anions is much less than that given to the determination of cations. This symposium reserved for the consideration of the determination of anions is indicative of an increasing awareness of the importance of this facet of the chemical analysis of water and the importance of having reliable methodology for the quantitation of the concentration of anions. The essential steps in the methodology involved in the determination of anions, namely: sampling; preliminary chemical treatments; analytical separations; and measurement of anion concentration by spectrophotometry, spectrofluorometry, and atomic absorption spectrometry will be discussed in sequence.

SAMPLING

As in most analytical determinations the total variance, V_T, depends on the variance of sampling S_s^2, the variance of preparation of sample, S_p^2, and the variance of the measurement.

$$V_T = S_s^2 + S_p^2 + S_m^2$$

The variance of sampling, S_s^2, should not exceed 50% of
the total variance, V_T.

It is essential that the water sample used for analysis
is truly representative of the natural, treated, or polluted
water being examined. From the viewpoint of both economy
and precision it is preferable to use a pooled (composite)
sample of 6 to 10 random grab samples and run several
replicate determinations. In every case a complete description of the sampling conditions should be recorded,
e. g., the temperature, local conditions, extent of turbidity,
odor, depth, distance from shore, rate of stream flow.

The type of containers used for sample storage and
the time permitted to elapse between collection of sample
and the beginning of the actual analysis are also important
considerations. Sulfide, sulfite and cyanide can be lost by
volatilization. Silicon and boron can be transferred to the
sample by leaching of the glass container. As will be
pointed out later, there are numerous different methods of
determining and reported phosphate in water. Depending
on the degree of differentiation desired, it may be necessary
to filter the sample immediately after collection. Some
evidence has been cited that phosphate is adsorbed by polyethylene bottles so that samples of very low phosphate
content should not be stored in plastic containers.

PRELIMINARY TREATMENTS

Acid hydrolysis to convert condensed phosphates to
orthophosphate, oxidative digestion to convert organically
bound phosphorus to orthophosphate, the fixation of sulfide
by addition of zinc acetate, and the removal of sulfide by
addition of lead carbonate to a basic (pH = 11) solution of
sample containing cyanide are typical examples of preliminary treatments. Preconcentration by the use of ion
exchange membrane papers, or by evaporation is sometimes
necessary.

ANALYTICAL SEPARATIONS

The nature and concentration of other ions present in water, often present in very much larger concentrations than the anion to be determined, must be considered in selecting a particular method. Analytical separations are often required to circumvent interference and improve reliability. Let us review briefly the role of analytical separations prior to the determination of anions. Table I summarizes typical separations of anions using the liquid-liquid extraction method.

Separation by volatilization processes is another method of importance in anion analysis. Derivatization or conversion to a volatile compound is often a necessary preliminary step. Table II lists representative examples of distillation methods used to separate specific nonmetals. A special quartz distillation apparatus is recommended for the distillation of methyl borate (1).

TABLE I

LIQUID-LIQUID EXTRACTION METHODS: ANIONS

D. C.*	Extraction System
CN^-	Fatty acids extr.; isooctane; pH 6-7
PO_4^{3-}	H_3P $(Mo_3O_{10})_4$; ether
ClO_4^-	$Cu(DMP)_2$, ClO_4; EtOAc
BF_4^-	Methylene Blue, BF_4; $C_2H_4Cl_2$
NO_3^-	$(C_6H_5)_4P$, NO_3; $CHCl_3$

*Desired Constituent

TABLE II

DISTILLATION METHODS: NONMETALS

D. C.	Volatilized As
CN^-	HCN
BO_3^{3-}	$B(OCH_3)_3$
F^-	SiF_4
AsO_4^{3-}	AsH_3
SeO_3^{2-}	$SeBr_4$
SiO_3^{2-}	SiF_4

Strong cation exchangers (SCX) and strong anion exchangers (SAX) can be used to partition cations and anions while weak anion exchangers (WAX) will partition conjugate bases of strong and weak acids. Table III cites typical ion exchange methods. The feasibility of separating phosphate species by ion exchange prior to the automation of the spectrophotometric method has been demonstrated (2, 3).

TABLE III

ION EXCHANGE METHODS: ANIONS

D. C.	Resin	Diverse Ions
BO_3^{3-}	SCX	$Ti(IV), Al, Fe(III)$
SiO_3^{2-}	SCS	$Fe(III), Al$
SiO_3^{2-}	WAX	$PO_4^{3-}, AsO_4^{3-}, CrO_4^{2-}$
PO_4^{3-}	SAX	$P_2O_7^{4-}, P_3O_5^{5-}$
F^-	SAX	Cl^-, Br^-
F^-	SAX	PO_4^{3-}
F^-	SAX	$Fe, Al; PO_4^{3-}, Cl^-, SO_4^{2-}$

Investigations have been made in which the thin layer chromatography method was utilized to partition various anions. Anions classified into Groups A, B, and C have been isolated by the TLC method (4). The halides have been partitioned using an ammoniacal mobile phase of acetone, n-butanol, and water (5).

METHODOLOGY IN THE DETERMINATION OF ANIONS

Indirect Methodology

In the spectrophotometric determination of nonmetals, it has often been necessary to resort to indirect methodology in order to develop satisfactory methods. Although a specific and sensitive direct spectrophotometric method is always preferred, the substitution of a well-delineated indirect method is acceptable. Most indirect methods involve complexation, precipitation and liquid-liquid extraction techniques.

Several classical colorimetric methods for the determination of fluoride are based on the decrease in color as the result of fluoride ions complexing with the zirconium(IV) or thorium(IV) initially present in a colored complex with a dye. Hence, the reduction of the absorbance of a zirconyl eriochrome cyanine R complex is proportional to fluoride (6). Fluoride forms a stronger complex with zirconium(IV) than the dye so that the decrease in color occurs as the fluoride removes the zirconium(IV) from the colored zirconium dye complex. Sulfide and cyanide have been determined by their decomplexation reactions with a palladium(II) chelate, bis(7-iodo-5-sulfoxino)-palladium(II). The chelating agent is generated in basic solution as $Pd(CN)_4^{-2}$ is formed. After acidification and addition of iron(III), the absorbance of the blue-green potassium tris(7-iodo-5-sulfoxinol iron(III) in presence of palladium(II) chelate) is measured (7).

There are two general methods in indirect spectrophotometric and atomic absorption spectrometric analysis.
Table IV delineates the nature of these two general methods.

TABLE IV

CLASSIFICATION OF INDIRECT ABSORPTION
SPECTROMETRIC METHODS

I.   FORMATION OF STOICHIOMETRIC COMPOUNDS
     CONTAINING DESIRED CONSTITUENT

          A. Determination of Equivalent Constituent
          B. Determination of Excess Reagent

II. DETERMINATION OF EFFECT OF DESIRED
     CONSTITUENT ON FORMATION OF ADSORPTIVE
     SYSTEM

          A. Diminution of Absorbance
          B. Enhancement of Absorbance

Several examples of Method I in which stoichiometric
complexes or precipitates are formed and the measurement of the equivalent constituent is the basis of
quantitation are given in Table V, the absorbance of the
displaced chloranilate being measurable in either the
visible or ultraviolet region.

Indirect atomic absorption spectrometric methods
of analysis have been presented previously (8).
Tables VI and VII indicate indirect A.A.S. methods
for anions based on Method IA and IB.

Another example of indirect methodology involving
Method IIA is the atomic absorption inhibition titration
(AAIT) technique recently reported by Looyenga and
Huber (9). This technique was applied to the determi-
nation of silicate in wastewater. In this technique the

TABLE V

INDIRECT SPECTROPHOTOMETRIC METHODS
FOR ANIONS

D. C.	Reagent	Wavelength, nm	Conc. Range, ppm
SO_4^{2-}	$Ba(C_6Cl_2O_4)_2$	530	20 - 400
		332	0.2 - 5
F^-	$Sr(C_6Cl_2O_4)_2$	332	5 - 50
PO_4^{3-}	$La(C_6Cl_2O_4)_2$	530	3 - 100
SO_3^{2-}	$Hg(C_6Cl_2O_4)_2$	525	6 - 125
		330	0.6 - 10
S^{2-}	$Hg(C_6Cl_2O_4)_2$	525	5 - 200
		330	0.1 - 1.5
CN^-	$Hg(C_6Cl_2O_4)_2$	525	20 - 200
		330	0.4 - 4

TABLE VI

INDIRECT A. A. S. METHODS: PRECIPITATION

D. C.	Precipitate	Element Detd.
Cl^-	$AgCl$	$Ag, Ag(NH_3)_2^+$
CN^-	$AgCN$	xs Ag
SO_4^{2-}	$PbSO_4$	xs Pb
	$BaSO_4$	$Ba, BaEDTA^{2-}$
SiO_3^{2-}	$(H_4SiMo_{12}O_{40})$	
	$PbMoO_4$	xs Pb

TABLE VII

INDIRECT A. A. S. METHODS: EXTRACTIONS

D. C.	Complex	Extractant	Element Detd.
ClO_4^-	$Cu(DMP)_2ClO_4$	$CHCl_3$; EtOAc	Cu
NO_3^-	$Cu(DMP)_2NO_3$	MIBK	Cu
$(C_6H_5)_4B^-$ K$^+$	$Cu(DMP)_2(C_6H_5)_4B$	EtOAc	Cu
SCN^-	$Cu(Py)_2(SCN)_2$	$CHCl_3$; EtOAc	Cu
CN^-	$Fe(Ph)_3(CN)_2$	$CHCl_3$; EtOAc	Fe
I^-	$Cd(Ph)_2I_2$	$C_6H_5NO_2$	Cd
Hg^{2+}	$Zn(BP)_2HgBr_4$	$1,2\text{-}C_2H_4Cl_2$	Zn
SO_4^{2-}	$Ba(ClA)_2$; $Fe(Ph)_3(ClA)_2$	$C_6H_5NO_2$	Fe

silicate inhibits the magnesium atomic absorption signal. Removal of cations (Mg^{++}) by cation exchange resin is first necessary. A standard magnesium(II) solution is titrated into sample solution which is then fed into an AAS burner. The endpoint is detected by increase in AAS signal for presence of magnesium added in excess of that required to titrate the silicate.

Indirect fluorometric methods are applicable to the determination of traces of ions which can not be converted readily to either a highly absorptive or fluorescent species. There are two general techniques available in indirect fluorometric methodology. In the first technique one or more stoichiometric reactions are used to produce a fluorescent compound whose fluorescence is proportional to the trace amount of a specific constituent participating in one of the reactions. Decomplexation reactions are frequently used advantageously in this analytical approach. For example, Hanker, Gelberg, and Witten developed an indirect fluorometric method for cyanide which was sensitive to 0.02 μg of cyanide per ml of solution by demasking of 8-hydroxy-5-quinolinesulfonic acid from the nonfluorescent potassium bis(5-sulfoxino)palladium(II) chelate and then complexing magnesium(II) with the liberated chelon to give a fluorescent potassium bis(5-sulfoxino) magnesium. A sensitive

indirect fluorometric method for the determination of sulfide
is based on a similar sequence of reactions (7).

The second technique employed in indirect fluorometric
methodology is often referred to as the "quenching method, "
in which the decrease of the fluorescence of a fluorescent
reagent is proportional to the amount of the constituent causing
the decrease. Hence, the stoichiometric reaction consists of
the conversion of a fluorescent reagent to a nonfluorescent
system. The indirect fluorometric determination of fluoride,
iodide, and hydrogen sulfide are based on "quenching" methods.
A summary of indirect fluorometric methods for certain anions
is given in Table VIII.

Selected Methods for Specific Anions

A number of anions of special interest in the analysis of
natural, treated and polluted waters will now be considered. In
selecting and evaluating a specific analytical method, the criteria
cited in Table IX should be considered.

TABLE VIII

INDIRECT FLUOROMETRIC METHODS
FOR ANIONS

D. C.	Reagent	Wavelength	Conc. Range
F^-	Zr-Flavanol	460	0.002 - 0.1 µg/ml
PO_4^{3-}	Al-Morin	510	1 - 6 µg P/ml
I^-	$UO_2(OAc)_2$	520	0.02 - 0.2 µg/ml
CN^-	K bis (5-sulfoxino) palladium(II): Mg(II)		0.1 - 0.8 µg/ml
S^{2-}			1 - 5 µg/ml
SO_4^{2-}	Zr-Calcein blue	350, 410	20 - 120 µg/ml
Citrate	Tungstate-Flavanol	450	0 - 0.2 ug/ml

TABLE IX

EVALUATION CRITERIA
FOR ANALYTICAL METHODS

1. Sensitivity
2. Specificity
3. Speed
4. Simplicity
5. Suitability for Automation
6. Precision and Accuracy
7. Cost
8. Applicability

Phosphate. Spectrophotometric, fluorometric and indirect atomic absorption methods are based on reactions involving the orthophosphate species. However, in natural waters and wastewaters the phosphorus may be present as orthophosphate, as condensed phosphates (pyro-, tripoly-, meta-) and organic phosphates. These phosphates are soluble but may also be present in suspended particles of earth or organisms.

There are many different classifications of phosphate in water based on chemical nature and physical states as indicated in Table X (10).

Acid hydrolysis converts condensed phosphates to orthophosphate. Organic phosphates require an oxidation-digestion to convert to orthophosphate.

In the spectrophotometric determination of orthophosphate, the precision and accuracy obtainable is dependent primarily on (1) method used, (2) concentration level, (3) optical path length of absorption cell. For low concentrations (less than 0.5 ppm) a heteropoly blue method is capable of better precision and accuracy. A percent relative standard deviation of 5 to 10% is reasonable as is a relative error of 10%.

TABLE X

CLASSIFICATION OF PHOSPHATE DETERMINATION
IN WATER

Chemical Nature	Total	Physical States	
		Filterable	Particulate
Total	Total dissolved and suspended phosphate	(dissolved) Total dissolved	Total particulate phosphate
Ortho	Total dissolved and suspended phosphate	Dissolved orthophosphate	Particulate ortho-phosphate
Acid-hydrolyzable	Total dissolved and suspended acid-hydrolyzable phosphate	Dissolved acid-hydrolyzable phosphate	Particulate acid-hydrolyzable phosphate
Organic	Total dissolved and suspended organic phosphate	Dissolved organic phosphate	Particulate organic phosphate

For the 1 to 5 ppm range, a relative deviation of 5% and a
relative error of 3-5% is possible. Only at the higher
phosphate (3-7 ppm) concentrations is the molybdovanado-
phosphoric acid method competitive in respect to accuracy
and precision (10). A variety of spectrophotometric methods
based on heteropoly chemistry are listed in Table XI.

TABLE XI

SPECTROPHOTOMETRIC METHODS: DETERMINATION OF PHOSPHATE

Method	Wavelength, nm	Conc. Range, ppm P
Molybdophosphoric Acid	400	5 - 25
Molybdophosphoric Extrn.	310	0.25 - 1.25
Molybdophosphoric IND.	230	0.1 - 0.55
Heteropoly Blue	830	0.2 - 1.2
Extrn. MPA - Het. Blue	725	0.25 - 1.25
Het. Blue - Extrn.	784	0.2 - 1.1
Molybdovanadophosphoric Acid	460	8 - 45
Mod. MVPA	315	0.3 - 1.5
Extrn. MVPA	315	0.25 - 1.25
Crystal Violet - MPA	582	0.05 - 0.50
MPA; 2-Amino-4-chloro-benzenethiol	710	0.025 - 0.1

Phosphate has been determined by an A.A.S. method involving the extraction of molybdophosphoric acid with 2-octanol,
the aspiration of the extract into an air-acetylene flame, and the
measurement of the absorbance at the molybdenum line (11).
n-Butyl acetate as extractant and a nitrous oxide-acetylene
flame have also been used in a similar procedure (12). Phosphate
and silicate (13, 14) and phosphate, silicate, and arsenate (15)
have been determined sequentially by the selective extraction of
molybdoheteropoly acids and atomic absorption spectrometry.

A recent publication reports a new spectrofluorometric method based on the formation of a fluorescent Rhodamine B molybdophosphate complex which is extractable into chloroform: butanol (16). An indirect method based on the quenching of the fluorescence of an aluminum-morin complex has also been reported (17).

Silicate. Spectrophotometric methods for the determination of silicate are based on the formation of 12-molybdosilicic acid. Direct methods utilize either the absorptivity of the yellow molybdosilicic acid or the near infrared absorptivity of the heteropoly blue produced by controlled reduction of the molybdosilicic acid (18). Indirect methods have been described in which the ultraviolet absorptivity of the molybdenum(VI) in the extracted molybdoheteropoly complex is used (19), or the equivalent molybdenum (12 Mo: 1 Si) is measured as peroxymolybdic acid (20). This equivalent molybdenum can also be measured as a 2-amino-4-chlorobenzenethiol complex (21) by a procedure similar to that developed by Djurkin, Kirkbright, and West (22) for the determination of phosphate. The relative optimum concentration ranges for the spectrophotometric methods for the determination of silicate are listed in Table XII.

TABLE XII

SPECTROPHOTOMETRIC METHODS:
DETERMINATION OF SILICATE

Method	Wavelength, nm	Conc. Range, ppm Si
Molybdosilicic Acid	350	1.8 - 13
Molybdosilicic Acid Extrn.	350	0.5 - 5
Heteropoly Blue	815	0.1 - 1.5
Heteropoly Blue Extrn.	775	0.1 - 1.2
INDIRECT - UV	230	0.06 - 0.5
IND. - MSA - Peroxy - MA	330	0.25 - 2.5
IND. - MSA - 2 - NH_2 - 4 - Cl - C_6H_5SH	715	0.05 - 0.2

An indirect atomic absorption spectrometric method
for silicate depends on the precipitation of the molybdate,
associated with the molybdosilicic acid, as lead molybdate
and measurement of the excess lead in the supernate by
A. A. S. (23).

Nitrite. The most extensively used spectrophotometric
method for the determination of nitrite is based on the Griess (24)
reaction which involves the diazotization of sulfanilic acid and
subsequent coupling with 1-naphthylamine. Numerous modifications have been suggested, primarily on the basis of changing
either the substance being diazotized or coupled (25-30). A
critical study of the diazotization and coupling reactions at the
microgram level by Rider and Mellon (31) delineated the optimum
conditions for maximum development of the azo dye using sulfanilic
acid and 1-naphthylamine. Although more sensitive spectrophotometric methods based on the formation of free radical
chromogens have been proposed by Sawicki et al (32), the Griess
reaction method remains the method of choice (33). Wada and
Hattori (34) have concentrated the azo dye on a strong anion
exchange resin column, eluted with 60% acetic acid and measured
the absorbance at 550 nm using 5.00 cm cells. By this methodology they were able to determine 1-100 nanograms of nitrite
nitrogen per liter of sea water. Several spectrophotometric
methods for the determination of nitrite are listed in Table XIII.

TABLE XIII

SPECTROPHOTOMETRIC METHODS:
DETERMINATION OF NITRITE

Method	Wavelength, nm	Conc. Range, ppm
Griess: Sulfanilic Acid 1-Naphthylamine	520	0.05 - 1.2
Antipyrine	343	1 - 6
UV; 4-Aminobenzenesulfonic acid	270	0.3 - 2.7
2,3-Naphthotriazole	355	1.5 - 8.5
1-Methyl-2-quinolone azine	520	0.01 - 0.13

Ultraviolet spectrophotometric methods for nitrite based on the measurement of the absorbance of diazotized 4-aminobenzenesulfonic acid (35) and of 4-nitrosoantipyrine (36, 37) have been proposed. 2, 3-Diaminonaphthalene reacts with nitrite to form 2, 3-napthotriazole which can be extracted with tetrachloroethylene and determined spectrophotometrically or fluorometrically (38).

Nitrate. Most spectrophotometric methods for the determination of nitrate are based either on the nitration of an organic compound, phenolic compounds in particular, or the oxidation of an organic compound. A few methods are based on the reduction to nitrite or ammonia (39, 40), or the formation of ion-association complexes. The phenoldisulfonic acid method is based on a nitration reaction (41) and the brucine method is presumably based on oxidation (42). Chromotropic acid (43) and Nile Blue A (44) have also been used to determine nitrate.

The characteristic high absorptivity of the nitrate ion at 203 nm has been used in the ultraviolet spectrophotometric determination of nitrate in water (45-47). The formation of nitrotoluene and extraction into toluene is the basis of an ultraviolet spectrophotometric method for nitrate, but can also be used for determining nitrites by employing a preliminary treatment with bromine (48).

Nitrate can be determined indirectly by extracting the ion-association complex, $Cu(2, 9\text{-dimethyl-}1, 10\text{-phenanthroline})_2 NO_3$, in methyl isobutyl ketone and measuring the copper by atomic absorption spectrometry (49). Tetraphenylphosphonium chloride in the presence of silver acetate gives an ion-association complex extractable with chloroform, the ultraviolet absorptivity being used in this case (50). These methods are summarized in Table XIV.

Sulfate. The turbidimetric method based on the formation of a barium sulfate suspension is still used in determining sulfate in water (51). A spectrophotometric method in which barium chloranilate is added to precipitate barium sulfate and release an equivalent amount of the hydrogen chloranilate ion is also available. The hydrogen chloranilate has a characteristic

TABLE XIV

SPECTROPHOTOMETRIC METHODS:
DETERMINATION OF NITRATE

Method	Wavelength, nm	Conc. Range, ppm NO_3^-
Phenoldisulfonic Acid	410	0. 1 - 2
Reduction: Griess	520	0. 06 - 1. 6
Ultraviolet	220	1. 5 - 15
Brucine	410	0. 4 - 4
Chromotropic Acid	410	0. 1 - 5
Toluene	284	1 - 18
Nitrato bis (2, 9- DM-1, 10 Ph) copper (I); (MIBK)	456	0. 5 - 4
Nile Blue A	650	2 - 20
Tetraphenylphosphonium Chloride	269	6 - 30

absorptivity in both the visible and ultraviolet regions (52). Another spectrophotometric method requires the reduction of the sulfate to sulfide with the subsequent formation of methylene blue by the reaction of hydrogen sulfide and p-amino-N, N-dimethylaniline in the presence of iron(III) chloride (53).

Sulfate has been determined by indirect atomic absorption spectrometric methods in which the excess barium ion remaining after precipitation was determined by measurement at 553.5 nm (54), or the isolated barium sulfate precipitate was dissolved in disodium dihydrogen ethylenediaminetetraacetate and the equivalent barium determined (55, 56). The precipitation of lead sulfate in an ethanolic solution and the determination of the excess lead by A. A. S. measurement at 283.3 nm has also been reported (57, 58). A spectrofluorometric method based on the fluorescence of a zirconium-calcein blue complex has been proposed for the determination of small amounts of sulfate ion (59).

Borate. For less than about 0.5 µg of boron per milliliter the curcumin method (60-63) seems to be applicable while for higher concentrations the carminic acid method is usually recommended (64-67). Both the curcumin (68) and carminic acid methods (69) have been automated. Quinalizarin (70,71) and 1,1'-dianthrimide (72,73) have been used as chromogenic agents for boron. Methylene blue (74), Brilliant green (75), methyl violet (76), and thionin derivatives (77-79) have been used to form colored ion-association complexes which are extractable with benzene or various chlorinated organic solvents, e.g., 1,2-dichloroethane. Table XV indicates the optimum concentration range for some of these spectrophotometric methods.

Borate and benzoin form a fluorescent boron-benzoin complex which serves as the basis of a fluorometric method for the determination of boron (80,81).

TABLE XV

SPECTROPHOTOMETRIC METHODS:
DETERMINATION OF BORATE

Method	Wavelength, nm	Conc. Range, ppm B
Curcumin (rosocyanin)	545	0.003 - 0.015
(rubrocurcumin)	540	0.01 - 0.10
Carminic Acid (H_2SO_4)	615	0.08 - 1.4
	300	0.04 - 0.4
(H_2SO_4-HOAc)	548	0.01 - 0.3
Quinalizarin (H_2SO_4)	610	0.04 - 0.5
(H_2SO_4-HOAc)	577	0.02 - 0.3
1,1'-Dianthrimide	635	0.02 - 0.4
Methylene Blue Tetra-fluoborate	660	0.01 - 0.16

Cyanide. The most extensively used spectrophotometric
method for cyanide employs the Konig reaction in which the
pyridine ring is opened by the action of cyanogen chloride and
the condensation of the resulting glutaconic aldehyde with 1-
phenyl-3-methyl-5-pyrazolone. Epstein (82) used chloramine T
to convert cyanide to cyanogen chloride and developed a stable
dye whose absorbance is proportional to the initial cyanide concentration. This pyridine-pyrazolone method was studied
critically by Kruse and Mellon (83). Improved sensitivity and
reproducibility can be achieved by extracting the colored dye
in butanol (84).

A spectrophotometric method based on the formation of extraction of the dicyano-tris(1,10-phenanthroline)-iron(II) complex
was developed by Schilt (85). An indirect atomic absorption
spectrometric method based on the measurement of the equivalent
iron in this complex has been reported (86). Lambert and Manzo
(87) have added the very insoluble tris(1,10-phenanthroline)iron(II)
triiodide to a cyanide solution with the release of the red tris
(1,10-phenanthroline) equivalent to the cyanide present. The
absorbance of the tris(1,10-phenanthroline)-iron(II) species is
measured. Cyanide has been determined spectrophotometrically
on the basis of the formation of a complex with mercury(II)-
methylthymol blue (88).

A specific fluorometric method for cyanide was developed by
Guilbault and Kramer (89). The fluorescence was produced by
the reaction of p-benzoquinone with cyanide. A catalytic spectrophotometric method using p-nitrobenzaldehyde and o-dinitrobenzene
has been reported (90). 5,5'-Dithiobis(2-nitrobenzoic acid) under-
goes a nucleophililic displacement reaction with cyanide with the
displaced thiol anion being measured spectrophotometrically (91).
4,4'-Dithiopyridine is more sensitive but absorbs at 324 nm.
Other indirect spectrophotometric methods involve inhibition
of the formation of a silver(I)-1,10-phenanthroline-Bromo-
pyrogallol red ternary complex (92) and the release of chloranilate
upon addition of mercury (II) chloranilate (93). The automation of
the spectrophotometric determination of cyanide in waters and
effluents has been described by Casapieri, Scott, and Simpson
(84). Table XVI summarizes several of the spectrophotometric
methods.

TABLE XVI

SPECTROPHOTOMETRIC METHODS: DETERMINATION OF CYANIDE

Method	Wavelength, nm	Conc. Range, ppm
Konig: Pyridine-Pyrazolone	620	0.04 - 0.2
Extrn Pyridine-Pyrazolone	630	0.04 - 0.2
Tris-(1,10-Phenanthroline)-iron(II) $(CHCl_3)$	597	1 - 8
Tris-(1-10-Phen)-iron(II) triiodide	514	1 - 8
4-4'-Dithiopyridine	324	0.3 - 1.3
Mercury(II); Methylthymol Blue	615	0.1 - 1.6
Catalytic: p-Nitrobenzaldehyde, o-Dinitrobenzene	560	0.045 - 0.45
INDIRECT: $Hg(C_6Cl_2O_2)_2$	525	20 - 200
	330	0.4 - 4
INDIRECT: Ag(I):; 1,10-Phen., Brompyrogallol Red	635	0.26 - 2.6

Halides. Chloride can be determined spectrophotometrically in which the absorbance of the chloranilate, released from mercury(II) chloranilate upon addition to a chloride solution, is measured (52). Another indirect method uses mercury(II) thiocyanate as a reagent to form the chloromercurate(II) anion and releases thiocyanate which reacts with iron(III) to produce a measurable color (95). In determining chloride in water, the thiocyanate released from the mercury(II) thiocyanate was extracted into nitrophenol as the tris (1,10-phenanthroline) iron(II) dithiocyanato complex (96). Chloride has been determined by an indirect atomic absorption spectrometric method in which silver chloride is precipitated, dissolved in an ammonia solution, and the equivalent silver determined (97).

Bromide in water has been determined spectrophotometrically on the basis of a preliminary oxidation to bromine
followed by the addition of either rosaniline (98), or phenol-
sulfonaphthalein (99).

Matthews and Riley (100) developed a spectrophotometric
method for differentiating the total iodide and iodate content of
sea water.

Although the fluoride ion-selective electrode method has
been extensively employed, the zirconium-xylenol orange method
has been found to be highly satisfactory for determining fluoride
in drinking water (101) and this method has been automated (102).

Anionic Surfactants. A commonly used biodegradable
synthetic detergent is an alkyl aryl sulfonate (LAS). An ion-
association complex of methylene blue - LAS extracted into
chloroform is the basis of a spectrophotometric method (103).
Anionic surfactants have also been determined by atomic
absorption spectrometry. The bis(1, 10-phenanthroline)copper(I)-
surfactant ion-associate is extracted into isobutyl acetate or
methyl isobutyl ketone and the equivalent copper determined
(104, 105).

CONCLUSIONS

Suitable spectrophotometric methods are available for the
determination of most of the anions commonly found in water.
The feasibility of automating spectrophotometric
is a distinct advantage. The main limitations are due to
interferences caused by diverse ions present in the water.
Spectrophotometric methods having improved selectivity are
desirable.

Atomic absorption spectrometry is applicable to the determination of anions primarily by employing indirect methodology.
In general, this technique is unsuitable for automation.

Spectrofluorometric methods are especially attractive from the viewpoint of the inherent sensitivity of this approach. Automation is usually possible. Quenching by foreign substances and the careful control of experimental parameters limits applicability in the determination of anions in water.

REFERENCES

1. Luke, C. L. and Flaschen, S. S. , Anal. Chem. , 30, 1406 (1958).

2. Lundgren, D. P. and Loeb, N. P. , Anal. Chem. , 33, 366 (1961).

3. Benz, C. and Paixo, L. M. , Chim. Anal. (Paris), 50, 247 (1968).

4. Kawanabe, K. , Takitani, S. , Miyazaki, M. , and Tamura, Z. , Bunseki Kagaku, 13, 976 (1964).

5. Seiler, H. and Kaffenburger, T. , Helv. Chim. Acta, 44, 1282 (1961).

6. Megregian, S. , Anal. Chem. , 26, 1161 (1954).

7. Hanker, J. S. , Gelberg, A. and Witten, B. , Anal. Chim. , 30, 93 (1958).

8. Boltz, D. F. , Developments in Applied Spectroscopy (Edited by E. L. Grove and A. J. Perkins) p. 222-6, Plenum Press, New York, 1969.

9. Looyenga, R. W. and Huber, C. O. , Anal. Chem. , 43, 498 (1971).

10. Am. Public Health Assoc. , Standard Methods for the Examination of Water and Wastewater, 13th Edition, p. 518-534, Washington, D. C. , 1971.

11. Zaugg, W. S. and Knox, R. J. , Anal. .Chem. , 38, 1759 (1966).

12. Kumamaru, T. , Otani, Y. , and Yamomoto, Y. , Bull. Chem. Soc. Japan, 40, 429 (1967).

13. Kirkbright, G. F. , Smith, A. M. , and West, T. S. , Analyst, 92, 411 (1967).

14. Hurford, T. R. and Boltz, D. F. , Anal. Chem. , 40, 379 (1968).

15. Ramaskrishna, R. V. , Robinson, J. W. , and West, P. W. , Anal. Chim. Acta, 45, 43 (1969).

16. Kirkbright, G. F. , Narayanaswamy, R. , and West, T. S. , Anal. Chem. , 43, 1434 (1971).

17. Land, D. B. and Edmonds, S. M. , Microchim. Acta, 1966, 1013.

18. Potter, G. V. , Colorimetric Determination of Nonmetals (Edited by D. F. Boltz), p. 47-74, Interscience, New York, 1958.

19. Trudell, L. A. and Boltz, D. F. , Anal. Chem. , 35, 2122 (1962).

20. Trudell, L. A. and Boltz, D. F. , Anal. Lett. , 3, 465 (1970).

21. Trudell, L. A. and Boltz, D. F. , Talanta, 19, 37 (1972).

22. Djurkin, V. , Kirkbright, G. F., and West, T. S. , Analyst, 91, 89 (1966).

23. Trudell, L. A. and Boltz, D. F. , Mikrochim. Acta, 1970, 1220.

24. Griess, P. , Ber. 12, 426 (1879).

25. Kieruczenko, A. , Chem. Anal. (Warsaw), 12, 103 (1967).

26. Warington, R. , J. Chem. Soc. , 39, 229 (1881).

27. Ilsovay, M. L. , Bull. soc. chim. , [3], 2 388 (1889).

28. Weston, R. S. , Proc. Am. Chem. Soc. , 27, 283 (1905).

29. Bratton, A. C. , Marshall, E. K. , Bobbitt, D. , and Hendrickson, A. R. , J. Biol. Chem. , 128, 539 (1939).

30. Barnes, H. and Folkard, A. R. , Analyst, 76, 599 (1951).

31. Rider, B. F. and Mellon, M. G. , Ind. Eng. Chem. , Anal. Ed. , 18, 96 (1946).

32. Sawicki, E. , Stanley, T. W. , Pfaf , J. , and Johnson, H. , Anal. Chem. , 35, 2183 (1963).

33. Am. Public Health Assoc. , Standard Methods for the Examination of Water and Wastewater, 13th Edition, p. 240-243, Washington, D. C. , 1971.

34. Wada, E. and Hattori, A. , Anal. Chim. Acta, 56, 233 (1971).

35. Pappenhagen, J. M. and Mellon, M. G. , Anal. Chem. , 25, 341 (1953).

36. Z. Perez, M. Acta Cient. Compostelana, 5, 159 (1968).

37. Weiss, K. G. , and Boltz, D. F. , Anal. Chim. Acta, 55, 77 (1971).

38. Wiersma, J. H. , Anal. Lett. , 3, 123 (1970).

39. Taras, M. J. , Colorimetric Determination of Nonmetals, Boltz, D. F. (Editor), p. 135, Interscience, New York, 1958.

40. Am. Public Health Assoc. , Standard Methods for the Examination of Water and Wastewater, 13th Ed. , p. 233, 454, Washington, D. C. , 1971.

41. Taras, M. J., Anal. Chem., 22, 1020 (1950).

42. Jenkins, D. and Medsker, L. L., Anal. Chem., 36, 610 (1964).

43. West, P. W. and Ramachandran, T. P., Anal. Chim. Acta, 35, 317 (1966).

44. Pokorny, G. and Likussar, W., Anal. Chim. Acta, 42, 253 (1968).

45. Hoather, R. C. and Rackham, R. F., Analyst, 84, 548 (1959).

46. Goldman, E, and Jacobs, R. J., J. Amer. Water Works Ass., 53, 187 (1961).

47. Armstrong, F. A. J., Anal. Chem., 35, 1292 (1963).

48. Bhatty, M. K. and Townshend, A., Anal. Chim. Acta, 56, 55 (1971).

49. Kumamatu, T., Otani, Y., and Yamamoto, Y., Bull. Chem. Soc. Japan, 40, 429 (1967).

50. Burns, D. T., Fogg, A. G., and Willcox, A., Mikrochim. Acta, 1971, 205.

51. Am. Public Health Assoc., Standard Methods for the Examination of Water and Wastewater, 13th Ed., p. 344, Washington, D. C., 1971.

52. Bertolacini, R. J. and Barney, J. E. II, Anal. Chem., 29, 281 (1957); 30, 202, 498 (1958).

53. Gustafsson, L., Talanta, 4, 227-243 (1960).

54. Dunk, R., Mostyn, R. A., and Hoare, H. C., Atomic Abs. Newsletter, 8, 79 (1968).

55. Borden, F. Y. , and McCormick, L. H. , Soil Sci. Soc. Amer. Proc. , 34, 705 (1970).

56. Roe, D. A. , Miller, P. S. , and Lutwak, L. , Anal. Biochem. , 15, 3 13 (1966).

57. Rose, S. A. , and Boltz, D. F. , Anal. Chim. Acta, 44, 239 (1969).

58. Little, I. P. , Reeve, R. , Proud, G. M. , Lulham, A. , J. Sci Food Agr. , 20, 673 (1969).

59. Tann, L. H. and West, T. S. , Analyst, 96, 281 (1971).

60. Bunton, N. G. and Tait, B. H. , J. Amer. Water Works Ass. , 61, 357 (1969).

61. Lishka, R. J. , J. Amer. Water Works Ass. , 53, 1517 (1961).

62. Spicer, G. S. and Strickland, J. D. H. , Anal. Chim. Acta, 18, 231 (1958).

63. Spicer, G. S. and Strickland, J. D. H. , J. Chem. Soc. , 1952, 4644-53.

64. Hatcher, J. T. and Wilcox, L. V. , Anal. Chem. , 22, 567 (1950).

65. Am. Public Health Assoc. , Standard Methods for the Examination of Water and Wastewater, 13th Ed. , p. 69, Washington, D. C. (1971).

66. Malyuga, D. P. , Agrokhimiya, 1969, 134.

67. Malyuga, D. P. , Zavod Lab. , 35, 279 (1969).

68. Hulthe, P. , Uppstrom, L. , and Ostling, G. , Anal. Chim. Acta, 51, 31 (1970).

69. Lionnel, L. J. , Analyst, 95, 194 (1970).

70. Gupta, H. K. L. and Boltz, D. F. , Mikrochim. Acta, 1971, 577.

71. Rudolph, G. A. and Flickinger, L. C. , Steel, 112, 114 (1943).

72. Gupta, H. K. L. and Boltz, D. F. , Anal. Lett. , 4, 161 (1971).

73. Danielsson, L. , Talanta, 3, 138 (1959).

74. Ducret, L. , Anal. Chim. Acta, 17, 213 (1957).

75. Babko, A. K. and Marchenko, P. V. , Zavod. Lab. , 26, 1202 (1960).

76. Poluektov, N. S. , Kononenko, L. I. , and Lauer, R. S. , Zh. Anal. Khim. , 13, 396 (1958).

77. Pasztor, L. C. and Bode, J. D. , Anal. Chim. Acta, 24, 467 (1961).

78. Pasztor, L. C. and Bode, J. D. , Anal. Chem. , 32, 1530 (1960).

79. Pasztor, L. C. , Bode, J. D. , and Fernando, Q. , Anal. Chem. , 32, 277 (1960).

80. White, C. E. , Weissler, A. , and Busker, D. , Anal. Chem. , 19, 802 (1947).

81. White, C. E. and Hoffman, D. E. , Anal. Chem. , 29, 1105 (1957).

82. Epstein, J. , Anal. Chem. , 19, 272 (1947).

83. Kruse, J. M. and Mellon, M. G. , Sewage and Ind. Wastes, 23, 1402 (1951).

84. Am. Public Health Assoc., Standard Methods for the Examination of Water and Wastewater, 13th Ed., p. 404, Washington, D. C., 1971.

85. Schilt, A., Anal. Chem., 30, 1409 (1958).

86. Danchik, R. S. and Boltz, D. F., Anal. Chim. Acta, 49, 567 (1970).

87. Lambert, J. L. and Manzo, D. J., Anal. Chem., 40, 1354 (1968).

88. Nomura, T., Bull. Chem. Soc. Japan, 41, 1619 (1968).

89. Guilbault, G. G. and Kramer, D. N., Anal. Chem., 37, 1395 (1965).

90. Guilbault, G. G. and Kramer, D. N., Anal. Chem., 38, 834 (1966).

91. Humphrey, R. E. and Hinze, W., Talanta, 18, 491 (1971).

92. Dagnal, R. M., El-Ghamry, M. T., and West, T. S., Talanta, 15, 107 (1968).

93. Humphrey, R. E. and Hinze, W., Anal. Chem., 43, 1100 (1971).

94. Casapieri, P., Scott, R., and Simpson, E. A., Anal. Chim. Acta, 49, 188 (1970).

95. Florence, T. M. and Farrar, Y. J., Anal. Chim. Acta, 54, 373 (1971).

96. Yamamoto, Y., Kumamaru, T., Tatehota, A., Yamada, N., Anal. Chim. Acta, 50, 433 (1970).

97. Westerlund-Helmerson, J., At. Absorption Newsletter, 5, 97 (1966).

98. Moldaln, B. and Zyka, J. , Microchem. J. , 13,
    357 (1968).

99. Archimbaud, M. and Bertrand, M. R. , Chim. Anal.
    (Paris), 52, 531 (1970).

100. Matthews, A. D. and Riley, J. P. , Anal. Chim. Acta,
    51, 295 (1970).

101. Macejunas, A. , J. Amer. Water Works Assoc. , 61,
    311 (1969).

102. Harwood, J. E. and Huyser, D. J. , Water Res. , 2,
    637 (1968).

103. Am. Public Health Assoc. , Standard Methods of Analysis
    for the Examination of Water and Wastewater, 13th Ed.,
    p. 339, Washington, D. C. (1971).

104. Courtot-Coupez, J. and LeBihan, A. , Anal. Lett. , 2,
    211 (1969).

105. LeBihan, A. and Courtot-Coupoz, J. , Bull. Soc. Chim.
    Fr. , 1, 406 (1970).

FLUORIDE ANALYSIS IN SEA WATER AND IN OTHER COMPLEX NATURAL
WATERS USING AN ION-SELECTIVE ELECTRODE--TECHNIQUES,
POTENTIALITIES, LIMITATIONS

Theodore B. Warner

Naval Research Laboratory

Washington, D. C. 20390

ABSTRACT

Fluoride activities and concentrations can be measured
simply and rapidly in a wide range of natural waters using
an ion-selective electrode. Direct potentiometry is convenient in sea water and appears well-suited for in situ
instruments. Known-addition techniques yield total concentration directly in sea water and in other fluids whose
ionic composition is not well-defined, such as water from
estuaries, rivers, or saline lakes, and no additional information about ionic strength or complexing ions is required. A combination of these methods provides information on the distribution of ions between free and complexed species. The general case is examined to define the
range of. solutions in which such methods may be used with
confidence. The principles apply equally well to electrodes
for other ions. Accuracy and precision is given for
measurements down to 0.02 mg F/Kg in a variety of solutions
containing different interfering substances, and a simple
approximate technique (good to a factor of 2) is discussed
for measurements in rain water down to 0.0003 mg/Kg (0.3
ppb).

INTRODUCTION

Ion-selective electrodes are well-suited for use in
complex natural fluids such as the water found in oceans,
estuaries, lakes, rivers, and rain, as well as many kinds

of biological solutions. Most electrodes are highly
selective and are useful down to total ion concentrations
of about 10^{-6} molar. Thus preliminary separation or concentration steps can often be eliminated and the measurement can be made directly on the sample as obtained.
Electrodes respond rapidly, and their response is a voltage
that is easily transmitted, stored, or measured. In
addition, the electrodes are simple and fairly rugged;
hence, they are promising for direct in situ measurements
in seawater (Warner, 1972). The main disadvantages of
using such electrodes directly are that accuracy is only
moderately good, response is temperature-dependent, and
other ions can interfere with measurements in a number of
ways.

This paper reviews several ways that fluoride concentrations can be measured in various complex natural
waters, using a lanthanum fluoride electrode, and compares
the advantages and disadvantages of each. Many of the
general principles discussed here will apply equally well
to other ion-selective electrodes, or to other related
aqueous solutions. The emphasis is on simplicity and
generality. How simply and rapidly can the measurements
be made? How wide a range of sample compositions can be
handled? How can complications be avoided?

Three methods are considered. The first is a direct
potentiometric technique specifically designed for use in
seawater and even more specifically aimed toward remote
sensing of fluoride concentrations with an in situ device,
knowing only cell voltage and salinity of the seawater
(Warner, 1969). Salinity is essentially a measure of the
total weight in grams of dissolved inorganic matter in
one kilogram of seawater. It is measured very quickly and
accurately by determining conductivity. The method described is only a partial solution to this overall goal of
in situ use because temperature variation has yet to be
dealt with; however, the method has proved useful in the
laboratory.

The second method is an adaptation of a known-addition
technique to the kinds of waters most often encountered in
the environment (Warner, 1971a). It too, is useful in seawater, but it also allows measurements to be made in a wide
range of complex solutions about which very little is known
regarding the concentrations of other interfering ions.

This includes water from estuaries, heavily polluted rivers, and saline lakes. The third method is a rough technique, good to a factor of 2, that can be used for measurements in rain or in other low-ionic-strength fluids down to about 0.3 parts per billion of fluoride.

POTENTIOMETRIC METHOD

When measuring fluoride total concentration in sea-water ($30 <$ salinity < 40) using a direct potentiometric method, the pH of the sample is fixed between 5 and 6 using an acetic acid-acetate buffer. The buffer is used in laboratory measurements to prevent drift in sample pH upon exposure to laboratory air. The electrode system is first standardized in Lyman and Fleming (1940) synthetic sea-water of known fluoride concentration C_O and salinity by determining cell potential E_s and evaluating B in the relation

$$E_s = B - S \ln C_o \qquad (1)$$

where $S = RT/F$ for the fluoride electrode, and R, T, and F are the gas constant, absolute temperature, and the faraday, respectively. Then cell potential E_x is measured in the unknown seawater sample, and the approximate total soluble fluoride concentration C_x is computed from the relation

$$E_x = B - S \ln C_x \qquad (2)$$

using the value of B determined in the standard. The salinity of the unknown is measured and actual total concentration of soluble fluoride C_s is computed from the relation

$$C_s = Q_a C_x \qquad (3)$$

where Q_a is a term that accounts for differences in the salinities, and hence the ionic compositions, of the synthetic standard and the unknown. The best experimental value (Warner, 1971b) for Q_a is

$$Q_a = \frac{\text{salinity}}{57.6} + 0.393 \qquad (4)$$

Theoretical and experimental values of Q_a are within about 1% of each other.

In practice, a potentiometric measurement of fluoride
in seawater takes about 15 minutes, and the coefficient of
variation is about 0.5% when measurements are made in a
thermostat aboard ship at sea. Total fluoride in seawater
is about 1.4 mg/Kg. We estimate that results may be biased
up to 1% due to small local variations in seawater composition.

Salinity differences between the unknown and standard
must be considered because, if the salinities differ, then
the chemical compositions differ and the fraction of the
fluoride present that is sensed as free fluoride activity
will also differ. In seawater, about half of the total
soluble fluoride C_s is present as free F^- ions, and half is
tied up in complexes, primarily with Mg^{++} and Ca^{++}.

$$C_s = [F^-] + [Mg F^+] + [Ca F^+] \quad (5)$$

The ratio of fluoride activity a_{F^-} and fluoride concentration $[F^-]$ is the activity coefficient γ,

$$\gamma = \frac{a_{F^-}}{[F^-]} \quad (6)$$

which depends on the total ionic strength of the solution
and is near 0.6 in seawater. The electrode responds only
to a_{F^-}. However, given the salinities of sample and
standard, one can make good estimates of ionic strengths
and of the fraction of the total fluoride that will be tied
up in calcium and magnesium complexes. Hence, the relation
between total ion concentration and activity can be calculated.

This technique was designed to be directly useful for
in situ measurements in the ocean. If the influence of
changing temperatures on the electrode system can be
minimized or compensated, then once the system is standardized only cell voltage and water salinity need be measured
to determine fluoride content. There is no need for sample
manipulation or addition of reagents, and hence it should
be possible to lower the sensors into the water rather than
to bring back discrete samples in a conventional oceanographic cast. A similar approach should be useful for
continuous unattended measurements in other complex
solutions.

Note that this measurement gives an estimate of free fluoride, and total concentration is obtained by calculation. This is important because the method discussed below measures total soluble fluoride directly, and hence a combination of both methods can give information on how much is free and how much is complexed.

KNOWN ADDITION METHOD

This technique includes the use of complexing buffers. It is useful for samples whose chemical composition is largely unknown, such as water from a heavily polluted river, a saline lake, or an estuary. In estuary water, for example, the ionic strength is high and varies over a wide range. Concentrations of Mg^{++} and Ca^{++} are high and vary over a wide range, thus hiding differing fractions of the total F^- by complexation. Furthermore, near river inputs or in rivers themselves there may be appreciable quantities of dissolved iron or aluminum, both of which complex fluoride very strongly.

Direct potentiometry is not possible in such waters unless one makes a number of auxiliary measurements to define the other variables. But measurements can be made quite simply using a known addition of fluoride and a complexing buffer. The buffer fixes pH in an optimum range, and includes cyclohexanediamine tetraacetic acid to tie up any iron and aluminum present. This complexing agent was originally suggested for river water by Harwood (1969). The cell potential is first measured in the buffered unknown. Then a known amount of fluoride is added directly to the sample and the new potential is measured. What this does, in effect, is to use the unknown sample as the chemical matrix for the standard as well. The standard and unknown will have the same ionic composition. Hence, the ratio between activity, which is what the electrode responds to, and total concentration, which is desired, is the same in sample and standard, and the difference in potentials before and after adding the spike yields total concentration directly. This is true even if the majority of the fluoride present is tied up in complexes.

The theory of the method has been well worked out by a number of investigators, but the technique has been

little used to date in complex natural fluids.   Recently,
Manahan (1970) used this approach to measure nitrate in
simple aqueous solutions, using a fluoride electrode as
reference, and Bruton (1971) used the method for determining fluoride and chloride simultaneously in fluorescent
lamp phosphors.   Our goal was to devise a technique that
could be used in virtually any surface water and also to
define the limits of generality.   That is, over what range
of values could interfering ion concentrations vary and
still not influence the accuracy of the final result.

The governing equation is

$$C_s = C_\Delta (e^{\Delta E/S} - 1)^{-1} \tag{7}$$

where:

C_Δ is a constant determined by the size of spike added
ΔE is the difference in observed potentials before
and after spiking
S is response slope of electrode.
Values of $(e^{\Delta E/S} - 1)^{-1}$ have been tabulated elsewhere as a
function of ΔE (Warner, 1971a).   In practice, one measures
ΔE, looks up the value of the exponential term, and obtains
fluoride concentration directly, regardless of how fluoride
is distributed between free and complexed forms, as long
as enough fluoride remains free to measure.   Furthermore,
the method requires no information or assumptions about
sample ionic strength.

But how general is this?   Where is the equation valid,
and where does it break down?   In the derivation there are
3 main assumptions:

1.   Changes in total ionic strength and sample volume
upon spike addition can be neglected.

2.   Electrode interferences (other than ions that can
complex fluoride) are not present in amounts that will
affect electrode response.

3.   The fraction ϕ of the ion being measured that is
free and uncomplexed remains essentially constant upon
spike addition.

The first two assumptions can be satisfied easily by
choosing suitable experimental conditions.   Assumption

three is also justified in most solutions. This is not
obvious, however, and the degree to which it will be
justified will depend on the amounts and kinds of complexing ions present in the sample.

In the presence of any given complex-forming ion, the
distribution of fluoride ions between the free and complexed forms will be governed by the concentration
equilibrium constant K_c for the formation of that particular complex. The fraction ϕ of the fluoride that is
free is given by

$$\phi = \frac{1}{1+K_c\left[M^{n+}\right]} \qquad (8)$$

where M^{n+} is the particular complexing ion under consideration.

When fluoride ions are added in the spike, some must
necessarily combine with M^{n+} ions and ϕ must increase
somewhat, since the concentration of free M^{n+} ions decreases. The question is: how much will ϕ change, and
under what conditions will this change be sufficiently
small so that the method may be used without significant
error?

If ϕ increases by 1% upon addition of a spike, this
would cause an error of about 2% in the calculated fluoride
concentration of the sample. Taking this as an upper limit
for the amount of error tolerable due to a change in the
fraction of free ions, then solution conditions can be
defined under which a spike technique can safely be used.

Assume that the amount of fluoride added in the spike
is about equal to the amount initially present in the
sample. This is the usual case. The resulting change in
ϕ will depend on the amounts and kinds of complexing ions
present, as outlined in Table 1. Most ions can be tolerated.
The only complex-forming ions that will invalidate the
assumptions of the method when it is used in naturally
occurring waters are those relatively few with formation
constants larger than 1000. For natural waters, this would
be primarily Al(III) and Fe(III). The concentrations of
these trivalent ions can be reduced to negligible amounts
by including cyclohexanediamine tetraacetic acid in the
buffer to complex them preferentially. Note in particular

Table 1.   Possible consequences of adding a fluoride spike
that doubles total fluoride concentration to a
solution containing complexing ions.

Since $\phi = \dfrac{1}{1+K_c[M^{n+}]}$

If	Then upon addition of the spike
$K_c[M^{n+}] < 0.01$	$\Delta\phi$ must be $< 1\%$
$K_c[M^{n+}] > 0.01$	three cases arise
a) if $[M^{n+}] \gg [F^-]$	$\Delta[M^{n+}]$ will be small and $\Delta\phi$ will also be small
b) if $[M^{n+}] \ll [F^-]$	trivial case; too little M^{n+} to interfere
c) if $[M^{n+}] \approx [F^-]$	$\Delta[M^{n+}]$ can be large; $\Delta\phi$ can exceed 1%

But in case (c):

Since $K_c[M^{n+}] > 0.01,$

and $[M^{n+}] \approx [F^-] \approx 10^{-5}M$ in natural waters,

then K_c must be $> 10^3$

that very large concentrations of ions with small formation
constants, such as calcium or magnesium, will cause no
difficulties.

In theory, then, a known-addition technique should be
well-suited to natural waters. To test the practical use
of the method, artificial waters were prepared simulating
water from estuaries, rivers, and saline lakes and containing much higher concentrations of Ca^{++}, Mg^{++}, Fe^{+++}, and
Al^{+++} than would normally be encountered in nature. In
estuary water whose salinity was from 7-35 $^o/oo$, the relative standard deviation was 1%; the method was essentially

unbiased and accurate to about 2%.

In river waters, measurements were made at fluoride
levels of 1 ppm and 0.1 ppm. At 1 ppm total fluoride, results were similar to those in estuary water. Again,
accuracy was about 2% regardless of interfering ion concentrations and precision was similar to that seen before.
In river waters containing 0.1 ppm fluoride, or less, slow
electrode response made measurements rather tedious, so a
modified technique was adopted that sacrificed accuracy for
speed. Results were biased some 10-25% and coefficients of
variation varied from 1 to 5%. A 20% error at this level is
20 parts per billion.

The results indicated that in almost any natural water,
measurements good to 0.02 parts per million can be made,
and a determination can be made on a sample with no prior
information or chemical pretreatment in about 30 minutes.
Very large concentrations of ions that form complexes with
low stability constants can be tolerated. For example,
good results were obtained in water containing 1/2 molar Mg,
but only 50 micromolar fluoride. This known-addition
technique also offers a convenient way to measure the amount
of fluoride impurities present in other reagents, because
commonly the impurity levels are highest in those salts
whose cations form fluoride complexes.

SLOPE METHOD

This technique is useful for very low fluoride levels.
For dilute samples where fluoride is below 10^{-6} molar, or
20 parts per billion, direct potentiometry becomes extremely tedious and a single measurement takes many hours
to make. Nevertheless, rain water samples containing from
0.3 to 20 ppb fluoride can be measured approximately by
merely comparing the rate of change of potential in the
unknown with several nearby standards. This is most conveniently done on a recorder -- a typical trace is shown in
Fig. 1. This is an expanded scale presentation, with most
of the cell potential bucked out with a voltage source. The
first segment gives the approximate voltage of the electrode in the unknown. The next shows that a 2 ppb solution
has less fluoride than the unknown, and the 4th segment
shows that a 5 ppb standard has more. Hence the unknown is
less than 5 ppb and more than 2 ppb in fluoride. A

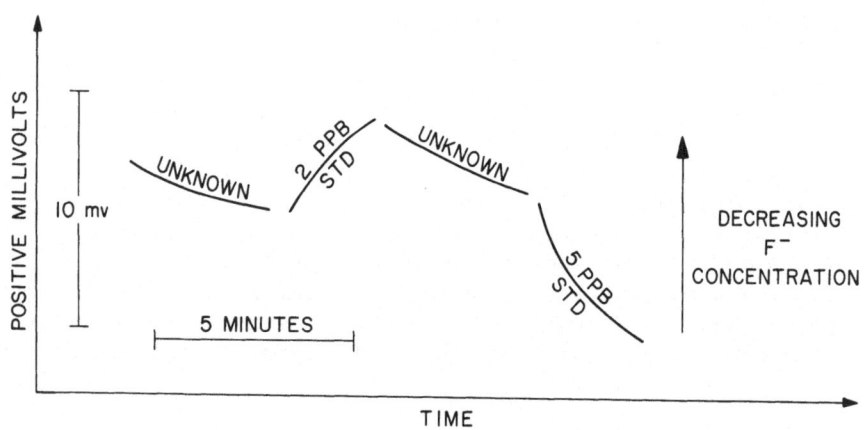

Fig. 1   Observed voltage changes when a rain water unknown containing about 3 ppb fluoride is alternated with
standards containing 2 and 5 ppb fluoride.

measurement takes about 15 minutes. In one test, a rain-
water sample was estimated to have 1.7 ppb fluoride using
this method -- it was then concentrated 10-fold by evaporation and remeasured and found originally to have had
1.5 ppb.

Recent results show that if acceptably fluoride-free
water can be obtained so that low-fluoride standards can be
prepared, then samples as dilute as 0.3 ppb can be reliably
measured with little increase in the time or effort required. Thus fluoride levels in even very pure rain, snow
or fog can be determined directly without need for preconcentration steps.

SUMMARY

Each of these three methods approaches the measurement
problem from a different point of view. Two of them are
essentially potentiometric, and hence measurements of free
ion activity. The third, based on a known addition gives

an estimate of total soluble fluoride without need for information about complexing ions present, changing liquid junction potentials, or ionic strengths. Thus total fluoride can be estimated one way, and the amount present in free, uncomplexed form in another. Because trivalent ions such as Fe or Al, which have complex formation constants well over 1000, tie up fluoride ions so tightly that they are essentially removed from solution, then known-addition measurements with and without the sequestering agent can give an estimate of just how much fluoride is tied up in these particular complexes. By exploiting the properties of these electrodes in varying ways, it is possible to build up a good picture of how a given substance is distributed among the various ionic forms available. This is most helpful in studies of natural water chemistry, or in studies of complex formation reactions, particularly in high ionic strength solutions which are frequently encountered in environmental studies. One can predict the various relationships between ion activity, free ion concentration, and total soluble element with surprisingly high accuracy. Thus measurements in other difficult systems should be relatively easy to make.

Which method is selected for finding total concentrations depends mainly on how well the chemical composition of the samples is known. The potentiometric technique is preferred where many analyses are to be made in well-defined samples, or where relative measurements will suffice. It is precise (coefficient of variation = 0.005 at the 1 ppm level) and rapid (5 to 15 min.). Known-addition determinations are less precise (C.V. = 0.010 to 0.015 at the 0.5 ppm level and slower (30 min.), but can be used in a much wider range of samples and are less prone to bias.

There are many other techniques available that have not been mentioned, such as potentiometric titrations, for example. These expand by an order of magnitude the amount of information one can extract from a given system, but they usually require more manipulation of the sample. Although fluoride has been emphasized here, all of the principles that apply to the fluoride measurement problem apply equally well to the use of any of the ion-selective electrodes. The ions and equilibria are different, but the principles are exactly the same.

ACKNOWLEDGEMENTS

I thank P. E. Wilkniss and D. J. Bressan for helpful editorial comments.

REFERENCES

Bruton, L. G. (1971), "Known addition ion selective electrode technique for simultaneously determining fluoride and chloride in calcium halophosphate," Anal. Chem. 43, 579-581.

Harwood, J. E. (1969), "The use of an ion-selective electrode for routine fluoride analyses on water samples," Water Research 3, 273-280.

Lyman, J. and R. H. Fleming (1940) "Composition of sea water," J. Mar. Res. 3, 134-146.

Manahan, S. E. (1970), "Fluoride electrode as a reference in the determination of nitrate ion," Anal. Chem. 42, 128-129.

Warner, T. B. (1969), "Fluoride in seawater: measurement with lanthanum fluoride electrode," Science 165, 178-180; a more complete description is given in Warner, T. B. (1969), "Measurement of fluoride activity and concentration in seawater using a lanthanum fluoride electrode," NRL Report 6905, 12 pp.

Warner, T. B. (1971a), "Electrode determination of fluoride in ill-characterized natural waters," Water Research 5, 459-465.

Warner, T. B. (1971b), "Normal fluoride content of seawater," Deep-Sea Research 18, 1255-1263.

Warner, T. B. (1972), "Ion selective electrodes - properties and uses in sea water," Marine Technology Society Journal 6, 24-33.

ANALYTICAL TECHNIQUES FOR THE DETECTION OF NUTRIENT ANIONS, PHOSPHATES AND NITRATES IN THE WATER ENVIRONMENT

Leonard L. Ciaccio [*]

GTE Laboratories, Inc.

Bayside, New York 11360

INTRODUCTION

Eutrophication refers to the natural aging processes of lakes. It carries two connotations: The increase in nutrients such as nitrates and phosphates leads to a greater biological productivity and eventually the conversion of a lake to dry land. The effects of eutrophication cause changes in plant and animal species present in the water body, which may be objectionable. The former development as a natural process is long term, although man's activities may greatly increase its rate of attainment. The latter result, however, is the undesirable effects, which may be the appearance of algal blooms and the loss of desirable fish species leading to destruction of recreational and drinking water resources (1). Nutrients and phosphates are responsible for algal blooms and their measurement in natural waters is important. The degradation and aging of rivers and streams, although not universally referred to as eutrophication, is considered a somewhat allied process and nutrients play a similar role as in the eutrophication of lakes and ponds (1).

A study of the nitrogen and phosphorus compounds in the environment indicates the complexity of the biogeochemical cycles in the ecosystem. The

* Present affiliation: Ramapo College of New Jersey, Mahwah

key to the mechanisms involved in eutrophication
is particularly dependent on the cycling of these
nitrogenous and phosphorus materials. The
chemical identification of these substances utilized by various organisms along with their
sources and rates of production need to be known
if a better understanding of eutrophication and
the correction of the over fertilization of our
water bodies is to be carried out (1).

PHOSPHORUS IN THE ENVIRONMENT

Sources

The sources of phosphorus arise from biological organisms and abiotic subtances in nature
and from the materials man has produced. The
weathering of minerals (see Table I) in the environment leads to soluble orthophosphates (see
Table II). Plants and animals on the other hand
yield organophosphates involved in energy transfer
in the body such as ATP, ADP, phosphoenolpyruvate
and phosphagens and inorganic phosphates such as
polymetaphosphates, pyrophosphates and orthophosphates. In addition, a host of other organophosphorus compounds arise from biological
organisms, e.g., phosphoglycerides, sphingo-
lipids, various enzymes, etc. (3). Man's effluents
contain inorganic condensed phosphates as pyrophosphates and tripolyphosphates, arising from
detergent products and poly and metaphosphates.
Pesticides composed of organic thiophosphates are
also present, although in low concentrations. The
large number of phosphorus-containing species
present in the dissolved and insoluble states are
indicated in Tables I and II. This portrays the
complexity of the phosphorus inputs to the
environment.

TABLE I

Solid Phase Forms of Phosphorus of Possible
Significance in Natural Water Systems[a]

Form	Representative Compounds or Substances
Soil and rock mineral phases:	
hydroxylapatite	$Ca_{10}(OH)_2(PO_4)_6$
brushite	$CaHPO_4 \cdot 2H_2O$
carbonate fluorapatite	$(Ca,H_2O)_{10}(F,OH)_2(PO_4,CO_3)_6$
variscite, strengite	$AlPO_4 \cdot 2H_2O, FePO_4 \cdot 2H_2O$
wavellite	$Al_3(OH)_3(PO_4)_2$
Mixed phases, solid solutions, sorbed species, etc.:	
clay-phosphate (e.g., kaolinite)	$[Si_2O_5Al_2(OH)_2 \cdot (PO_4)]$
metal hydroxide-phosphate	$[Fe(OH)_x(PO_4)_{1-x/3}],$ $[Al(OH)_x(PO_4)_{1-x/3}]$
clay-organophosphate	$[Si_2O_5Al_2(OH)_4 \cdot ROP],$ clay-pesticide, etc.
metal hydroxide-inositol phosphate	$[Fe(OH)_3 \cdot$ inositol hexaphosphate]
Suspended or insoluble organic phosphorus:	
bacterial cell material	Inositol hexaphosphate or
plankton material	phytin, phospholipid,
plant debris	phosphoprotein, nucleic
proteins	acids, polysaccharide phosphate

[a]Reprinted from Ref. (2) by courtesy of Wiley-
Interscience, N.Y.

TABLE II

Dissolved Phosphorus Forms of Possible
Significance in Natural Waters[a]

Form	Representative Compounds or Species
Orthophosphate	$H_2PO_4^-$, HPO_4^{2-}, PO_3^{3-}, $FeHPO_4^+$, $CaH_2PO_4^+$
Inorganic condensed phosphates:	
pyrophosphate	$H_2P_2O_7^{2-}$, $HP_2O_7^{3-}$, $CaP_2O_7^{2-}$, $MnP_2O_7^{2-}$
tripolyphosphate	$H_2P_3O_{10}^{3-}$, $HP_3O_{10}^{4-}$, $P_3O_{10}^{5-}$, $CaP_3O_{10}^{3-}$
trimetaphosphate	$HP_3O_9^{2-}$, $P_3O_9^{3-}$, $CaP_3O_9^-$
Organic orthophosphates	
sugar phosphates	Glucose-1-phosphate, adenosine monophosphate
inositol phosphates	Inositol monophosphate, inositol hexaphosphate
phospholipids	Glycerophosphate, phosphatidic acids, phosphatidyl choline
phosphoamides phosphoproteins	Phosphocreatine, phospho-arginine
Organic condensed phosphates	Adenosine-5′-triphosphate, coenzyme A
Phosphorus-containing pesticides	$O_2N\langle\bigcirc\rangle OPS(OCH_3)_2$; etc.

[a]Reprinted from Ref. (2) by courtesy of Wiley-
Interscience, N. Y.

The levels of total phosphate can vary from
<1 µg/liter to about 2 mg/liter. Some average
values would be between 13 and 38 µg/liter (4).
However, besides the large number of phosphorus·
containing substances which may be present, the
physical state of the phosphorus— bearing substances
also contribute to the complexity of the situation.
Four physical states which should be considered are
soluble, insoluble, colloidal and adsorbed states.
The adsorption/desorption activity of sediment and
dissolved phosphorus-containing substances (5)
indicates the phosphorus storage function of
sediments and rapid availability of phosphorus-
containing substances through this mechanism (6).
The role of phosphorus compounds in various
physical and chemical states is not fully understood although it has been shown that zooplankton
and metazoa digest particulate matter to obtain
phosphorus but do not assimilate soluble inorganic
and organic phosphorus compounds (7), while
aquatic plants assimilate inorganic phosphorus
forms but not organic phosphates (8). The role of
the microorganism is to carry out mineralization
thus converting organic to inorganic phosphates
(8,9). Thus in the consideration of an analytical
scheme both chemical and physical forms of
phosphorus compounds should be discernible.

Analysis

In the analysis for phosphates in environmental
water samples, the presence of a number of phases
and chemical forms of phosphorus complicate the
analysis scheme. Many investigators have en-
deavored to make some logical separations for the
physical forms and various chemical species.
Olsen (10) has prepared a broad review for the
analysis of orthophosphate in water samples and has
illustrated the various schemes by most of the investigators in this field.

Considering the physical entities present in
the water environment, one can make a division into
four following types of material: undissolved
(non-colloidal), colloidal, adsorbed and dissolved.
Undissolved (non-colloidal) matter consists of

seston, benthic sediments, nekton and pleuston.
Seston is composed of particulate matter containing living materials as plankton and dead matter
and tripton (detritus of an inorganic nature).
Organic and inorganic detritus, minerals and
organisms constitute benthic sediments. Nekton
and pleuston are swimming or free floating
aquatic plants, respectively, and are considered
a separate entity and not a direct supplier of
available phosphorus   Colloidal materials can
consist of organic and inorganic matter. Adsorbed
organic and inorganic phosphorus compounds appear
to be associated with surfaces of living organisms
such as nekton and pleuston and possible some
minerals. Dissolved matter is another type of
material which is considered when analyzing for
phosphates in the environment.

Thus, in the category of physical forms,
particulate matter as represented by undissolved
and colloidal matter, adsorbed species and dissolved
substances, are considered in various analytical
schemes. The adsorbed fraction has been considered
important for quite some time (11) since it would
provide a ready source of phosphorus compounds.
However, only recently has a technique been
suggested whereby this fraction can be recovered
from sediments (12).

Chemical forms of phosphorus in the environment include inorganic, ortho and polyphosphates and organic ortho and polyphosphates,
referred to in some schemes as residual phosphates
(10).

A number of separation schemes have been
proposed which utilize a combination of the physical
and chemical characteristics to isolate specific
fractions. A recent proposal by Olson (10) has
presented a fairly detailed scheme using both the
physical and chemical categories, i.e., dissolved,
colloidal and sestonic phosphorus- containing
phases and total orthophosphates, total polyphosphates and total residual (organic) phosphates.
Combining these three physical and chemical
categories, nine fractions theoretically can be
separated.

The problems in separating sestonic
(suspended) particulate matter and colloidal matter
complicate the issue. If sufficient care is taken,
the dissolved phase can be obtained with a separation of colloidal and suspended substances.
However, the usual practice of using a 0.45 μm
filter which would theoretically retain the
sestonic material but allow the colloidal matter to
pass through, does not function in this manner. Due
to the electrical charges on the colloidal matter
and the millipore filter, the colloidal matter is
retained with the sestonic materials (10). Because
of these practical limitations in the separation of
colloidal and sestonic material, this goal of
complete resolution of physical fractions cannot
be achieved at the present time. The use of silver
filters has been suggested by Olsen (10). Another
possibility is the use of continuous flow and
density gradient centrifugation to make this
separation (13). For the recovery of phosphorus
in the adsorbed phase, Williams et al (12) have
used electrodialysis (14) on sediment samples.

Procedures and schemes have been employed by
a number of investigators. A number of them will
be compiled here along with comments respecting
their shortcomings. The unfortunate situation is
that many of the values in the current literature
are called into question because of the inefficiency
and overlap of the physical and chemical separation
procedures, doubt with respect to completeness of
detection by the chemical method and the role of
interferences and their effect on the analysis.
Some of these analyses will be discussed, and if
the weak points of these separation schemes and
analyses are pointed out, perhaps more care will
be taken by future investigators.

The steps undergone in analyzing a sample
consist as follows: Physical fractionation,
chemical differentiation, conversion of the
phosphorus substance to orthophosphate and
analysis of the orthophosphate present. Several
of these steps are combined at times in one
operation.

In Table III below, the physical fractionation processes are given:

TABLE III

Physical Fractionation

Filtration thru 0.45 μ filter (15)

    Filtrate - soluble P

    Residue - insoluble P

Electrodialysis (12)

    Adsorbed P on sediments
    (hydrated iron compounds)

Dilute $HCl-H_2SO_4$ extraction (16)

    Adsorbed P, fluoapatite 88% and
    calcium phytate (11%)

Silver filters (10) or centrifugation (13)
    Colloidal P, suggested techniques

The errors inherent in filtration were
mentioned above regarding colloidal matter (10).
Thus, depending on the electrical state of the
filter residue, colloidal matter may be distributed between the filtrate and residue. The
imperfect resolution of the acidic extraction is
also indicated. It is included here because it
has been utilized for adsorbed P which does
present some problems in recovery, but nevertheless is an important source of available orthophosphate (16).

In Table IV techniques for chemical
differentiation and conversion of phosphorus
forms to orthophosphate as applied to the various
physical forms isolated are given.

TABLE IV

Chemical Differentiation

(In all these procedures all forms of organic and inorganic phosphorus are converted to orthophosphate.)

Orthophosphate content, PO_4

Use sample as received

Inorganic mineral phosphate (fluoapatite) P_{im}

Dilute HCl - H_2SO_4 extraction (16,17) (pH 1.1), P_{AE}

Inorganic, condensed phosphate, P_{ic}

Acid digestion (15), P_{AD}

Organic phosphate, P_o

Acid digestion in presence of an oxidizing agent, P_{AO}

Perchloric acid (18) and sulfuric-nitric acids (18) give good conversion to orthophosphates, while 30% H_2O_2 (15) does not give consistently good results. Persulfate works extremely well for estuarine waters (15,19).

The computations and limitations of the calculated species such as inorganic mineral phosphates, P_{im}, organic phosphorus, P_o, and inorganic condensed phosphates, P_{ic}, are illustrated in Table V.

TABLE V

Calculation of Chemical Species[a]

Orthophosphate, PO_4

   Analysis of sample without any treatment

Inorganic mineral phosphate, P_{im}

   $P_{im} = P_{AE} - PO_4$
   (Some digestion of condensed phosphates
   may occur.)

Inorganic condensed phosphates, P_{ic}

   $P_{ic} = P_{AD} - P_{AE}$
   (Some organic phosphorus may hydrolyze.)

Organic phosphorus, P_o

   $P_o = P_{AO}$ (total phosphate) $- P_{AD}$

   The error in P_{AD} due to some previous
   organic phosphorus hydrolysis could
   make P_o low

[a]The values are obtained indirectly except
 for orthophosphate present (PO_4)

Thus Tables III through V indicate the problems inherent in the physical and chemical differentiation techniques. The uncertainties introduced by the lack of complete resolution and chemical differentiation lend some doubt about the accuracy and reproduceability of reported values. Also, since conditions and presence of other species certainly affect the test, the bias in the analytical values will include this perturbation.

Another equally important problem concerns the relation of the physical process to the

process as it occurs in nature. Does the recovery of orthophosphate in the electrodialysis
process relate to the supply of phosphate from
the desorbing processes in nature? What sizes
of particulate matter are critical in nature?
Our efforts at differentiation must be rooted to
the ecological importance of the chemical and
physical form. A lack of information prevents the
answering of these questions fully. However, this
aspect should be kept in mind when analytical and
physical chemical techniques are being planned
and designed.

Since the main analytical detection procedure
is to convert all forms of phosphorus compounds to
orthophosphates, the requirement has been for a
fast and accurate method for the orthophosphate ion
devoid of interferences by other species. An exception is where resolution and identification of
organophosphorus substances is desired and this
problem will be discussed later. Olson has
adequately reviewed the genesis of orthophosphate
methodology for water samples and the ideal
characteristics of an orthophosphate method. They
are: specificity allowing distinction from inorganic condensed phosphates, P_{ic} (see Tables IV
and V), organic phosphates and other elements and
chemical species, a large Beer's law range with
sensitivity down to 0.01 µg/l with good accuracy,
and the absence of temperature and salt effects (10).
A review of several orthophosphate methods with
pertinent references is given below.

The most effort has been expended on the
molybdenum blue method whose basis is the formation
of the heteropoly complex, phosphomolybdate, with
subsequent reduction to phosphomolybdenum blue.
The reduction step has been the point of some
modification of the procedure. The reductant,
stannous chloride (20), was replaced by ascorbic
acid (21) which provided speed and convenience (22).
The addition of potassium antimony tartrate to the
ascorbic acid reductant gave yet another dimension
to the analysis (23). This development by Murphy
and Riley (23) is based on Jean's observation that

some elements as Ti, Zr or Bi, which react with the
phosphorus atom, catalyzed the formation of
molybdenum blue (24). The method of Murphy and
Riley (23) has some advantages when compared to
other methods as seen in Table VI. It is a sensitive test, uses one reagent, shows a low temperature and salt error and has a good tolerance for
some metals although it responds to arsenate.
Johnson used this response to arsenate to develop
a combined procedure for phosphate and arsenate (25).
The phosphomolybdate complex can be extracted by
organic solvents and the molybdenum blue color
developed in the solvent phase (10,15,19,29). The
advantage of this separation step is the elimination
of most interferences and an increase in method
sensitivity (30). The extraction method was extended to provide a very sensitive analytical
technique in that the concentration of the extracted molybdenum is determined using thiocyanate
as a reagent. This leads to the ability to detect
0.02 µg of phosphate with a 5% accuracy (31,32).
The choice of a particular orthophosphate analytical
technique depends on a number of factors and must be
determined for the sample under consideration.

It should be pointed out that the techniques
given in Table VI should be applicable in the use
of automated analytical instrumentation. An application is seen in the work of Lundgren (33). The
extraction method can also be carried out in such
equipment (34).

A recent publication gives a gas chromatographic procedure for the determination of a number
of anions. Phosphate, carbonate,sulfate and arsenate
are converted to trimethylsilyl derivatives. No
sensitivities have been reported, however. The
procedure allows for a determination of these anions
in the presence of one another. The peaks which are
sequentially eluted from a 12 ft. 3% OV-17 on a gas
chrom Q column are detected by flame ionization.
This technique should prove very useful in supplying
a more comprehensive and sensitive technique for the
analysis of anions in environmental samples (35).

TABLE VI

Compilation of Orthophosphate Methods

Method	Characteristics	Time Min.	λmax nm	Sensitivity[b] A/(μg/ml)/cm
Mo Blue ($SnCl_2$)	Two reagents a. molybdate b. stannous chloride	10	690	0.80
Mo Blue (Murphy & Riley)	One reagent a. molybdate b. Sb catalyst c. ascorbic acid reduction	10	882	0.58
Mo Blue (modification of Ref 23)	a. reduction b. Mo blue (Murphy & Riley)	40	720	0.52
Mo Blue (analysis for PO_4 and AsO_4)	a. Total PO_4+AsO_4	90	865	0.55
	b. Reduction PO_4	(15)[a]	865[a]	(0.26)[a]
	c. Difference AsO_4			
Vanadomolybdate	Vanadamolybdate reagent	20	420	0.06

TABLE VI (Continued)

Method	Tolerance	Interferences	Ref.
Mo Blue (SnCl$_2$)	Temp & salt errors	AsO$_4^{3-}$, WO$_4^{3-}$, SiO$_4^{4-}$, Ge^{4+}, Cr^{6+}, NO$_2^-$	(29)
Mo Blue (Murphy & Riley)	Low salt & temp. error Cu^{2+}, Fe^{3+}, SiO$_4^{4-}$	Sn^{2+}, Cr^{6+}, NO$_2^-$, AsO$_4^{3-}$	(23) (26)
Mo Blue (modification of Ref 23)	Low salt & temp. error AsO$_4^{3-}$, Cu^{2+}, Hg^{2+}, SiO$_4^{4-}$, Ce, Bi, SCN$^-$, Fe^{3+}, V, Au^{3+}, Ti^{3+}, Sn, Ge^{4+}	WO$_4^{3-}$	(27)
Mo Blue (analysis for PO$_4$ and AsO$_4$)	Similar to those above	Sn^{2+}, WO$_4^{3-}$	(25)
Vanadomolybdate	AsO$_4^{3-}$ (<125 ppm), Fe^{2+} (<100 ppm)	S^{2-}, SCN$^-$, F$^-$, Bi	(16) (17) (28) (29)

^{a}Arsenate analysis

^{b}A/(μg/ml)/cm ≡ Absorbance/((μg of PO$_4$-P/ml)/cm of cell thickness

The large number of organic phosphorus substances in the water environment present in a
variety of physical forms present the problem of
chemical identification and designation of their
part in the ecological system. Subfractionation
into groups depending on their utilization by the
various organisms is an ultimate goal in order to
shed light on energy flows and material cycling.

Christman and Minear (36) examined the use of
Sephadex G15 and G25 in separating some soluble
phosphorus compounds such as sodium glycero-
phosphate, calcium magnesium phytate, DNA and
potassium dihydrogen phosphate (see Table II).
They then took aqueous extracts from freeze dried
solids obtained from the filtered (0.45 μ membrane
filter) batch cultures of <u>Chlamydamonas Reinhardtii</u>
grown in synthetic medium and subjected them to
Sephadex chromatography. Chromatographic peaks
representing inorganic orthophosphates, organic
phosphorus substances and large molecular weight
organic phosphorus compounds were obtained. The
latter compound shows uv absorption suggestive of
the presence of nucleic acid material. They also
review briefly and summarize the literature concerning soluble organic phosphorus substances
found in the environment and give a figure to
illustrate the cycling of these materials (Fig. 8
in Ref. 36).

In the last analysis the resolution and
designation of the variety of useful fractions of
phosphate-bearing materials depends both on a
biological and geochemical function. The phosphorus entities in the phosphorus biogeochemical
cycle and other cycles such as calcium and iron,
influenced by phosphorus species indicate chemical
and physical species which should be considered.
However, since the complete nature of the cycle is
not known, judgment based on presently known biological, chemical and physical information must be
used. Refinement of our knowledge depends on the
development of suitable separation and analytical
techniques reflective of the operational modes
inherent in the environment. In addition a constant re-evaluation of the data used to explain

biogeochemical cycles must be continued in order
to determine the relevancy of the methods and
techniques used to obtain the data.

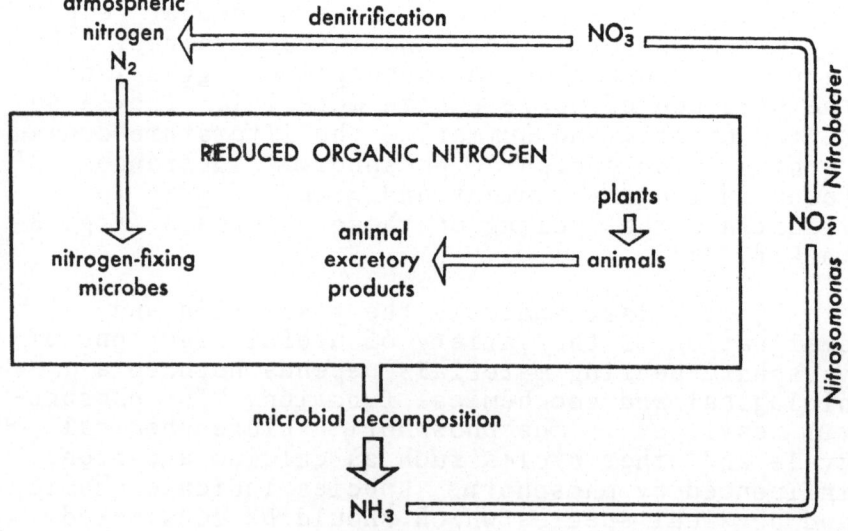

Figure 1. The nitrogen cycle. The interconversions
between the various forms of organic nitrogen (plant,
animal, and microbial) are shown in barest outline.

Reprinted from Ref. (37) by courtesy of Holt, Rinehart
and Winston, Inc., New York

NITROGEN IN THE ENVIRONMENT

Sources

The major sources of nitrogenous compounds are dead plant and animal matter, excretory products of animals, the atmosphere, including rain, and to a much smaller extent, minerals (37,38,39).

In Figure 1, a simple diagram of the nitrogen cycle illustrates the pathways of nitrogen exchange (37). Man has perturbed the distribution of nitrogen in the environment because of his agricultural, industrial and domestic activities (40) and the resulting degradation of our water supplies calls for a solution to this problem.

Water resources vary in their content of nitrogenous substances. The nitrate content of several rivers as measured in 1961 varied from 0.5 to 4.7 mg/1 and for a number of wells from 0.0 to 1.4 mg/1 (41). Values for some lakes are: NH_3-N, 49-544 µg/1 and NO_3^--N, 10-175 µg/1 (42).

Man's activities supply nitrogenous substances to nature water bodies. Painter (43) has recently reviewed the chemical content of wastes and water effluents. The NH_3-N content (ammonia calculated as nitrogen) varied from 41-53mg/1, with an average of 21 mg/1 for one U.K. town, while for a town in the U.S. this value varied from 4 to 35 mg/1. He also reports on the organic nitrogen content of wastes. Lee gave an estimate of tens of mg/1 of N due to nitrogenous substances in waste waters (44). In the corn belt of the U.S. fertilizer nitrogen is responsible for 55 to 60% of the NO_3-N found in surface water (45). The Sangamon River, thirty-four miles above Lake Decatur which provides drinking water for the city of Decatur, Illinois, showed a dramatic increase in nitrate content. For the years 1956 to 1961 the median value of NO_3-N was 2 mg/1, while for the period 1966 to 1969 the value was 7.4 mg/1. This can be attributed to farm runoff. The value is quite close to the acceptable PHS standard for drinking water - 11 mg of NO_3-N/1 (46).

Drainage water from irrigated fields show an increase in NO_3-N content (47). While the applied
water contained 0.25 mg/l, the subsurface drain
water contained 2.5 mg/l and the surface drain water
0.8 mg/l for the year 1959-60 in the Yakima Valley,
Washington (47).

The degradation of water resources can lead
to a number of problems besides eutrophication.
The presence of nitrates and nitrites in the environment leads to some health problems when considering drinking water. Nitrite will react with
hemoglobin to form methemoglobin. This latter
compound is a brownish material in which the iron has
been oxidized to the ferric state, in contrast to
hemoglobin where iron is present in the ferrous
state. The methemoglobin is not capable of picking
up oxygen in the lung and transporting it to the
place of need in the body. The vapor pressure of
oxygen in the lung is too low for this compound to
function as an oxygen carrier. However, nitrite is
not usually present to any large extent in the environment since the tendency is for it to be converted to nitrate. Nitrate, however, presents a
problem, especially in infants. In their digestive
tracts there are bacteria that convert nitrates to
nitrites. Thus, in vivo formation presents a great
hazard in children of this age. It has been noted
that a number of infant deaths have resulted from
the use of well water in Minnesota (48). For this
reason the Public Health Service has placed a limit
of 11 mg of NO_3-N/l in drinking water (46).

Thus, the sources of nitrogenous materials along
with the impact both on man's health and the state
of the environment indicate the breadth of consideration which should be given to the presence of
nitrogenous substances.

Analysis

Nitrogenous compounds are present in the environment in a number of chemical and physical forms.
Inorganic species include nitrates, nitrites and
ammonia or ammonium salts. Organic nitrogenous
substances include a host of compounds derivable

from living substances such as proteinaceous matter
(amino acids, enzymes, proteins, peptides, etc.),
nucleic acids and their fragments (purine and
pyrimidine bases), vitamins (thiamine, riboflavin,
nicotinic acids, etc.), alkaloids and other substances.

The inorganic forms, nitrates, nitrites and
ammonium ions are very soluble and are present as
ionic moieties. Some of the organic forms are insoluble and are present in sediments in lakes and
rivers (49,50). The distribution of nitrogenous
substances in sediments approximately mimics that
of soil as claimed by Lee (7) and is largely organic
nitrogen, about 98% of total nitrogen content. The
organic nitrogen content of soils is distributed as
follows: 20 to 50% as bound amino acids, 10% as
purine and pyrimidine derivatives and 5 to 10% as
combined hexoseamines (7). The considerations
discussed above for phosphorus containing substances regarding the separation of physical and
chemical forms are similar for nitrogenous materials
found in the environment. No further comment will
be made about separation of various physical forms
since the considerations are similar.

TABLE VII

Lake Mendota Nitrogen Content[a,b]

Constituent	Mean	Extremes
Inorganic	µg/liter	
N - NH_3	132	68 - 205
N - NO_3	32	10 - 72
Organic[c]		
Plankton	103	39 - 242
Total Soluble	593	377 - 969
Phospho-tungstic acid ppt (amino-N)	170	113 - 337
Free Amino	88	36 - 236
Peptides	181	77 - 436
Organic Nonamino	187	96 - 278

[a]See Refs. (42) and (51).
[b]The organic and inorganic values were for samples
taken at different times.
[c]Surface water samples

The distinction of inorganic and organic forms
of nitrogen in a lake is illustrated in Table VII.

The analyses reported in Table VII were
carried out in the 1920's so that the organic
analyses reflect the state of the art at that time.
However it does illustrate the variety and range
of concentrations of a number of fractions of nitrogenous substances.

Chemical and physical differentiations are made
through a number of operational categories which are
listed below in Table VIII.

The analyses necessary to determine the soluble
and suspended organic nitrogen fractions are given in
Table IX. The methods for ammonia nitrogen include
the classical distillation and Nesslerization procedure (15) along with one which allows for the
oxidation of ammonia to nitrite (Ref. 52 and Table
XI). From the comments given in Table IX with respect to interferences, an idea of the uncertainties
in the organic nitrogen fractions listed in Table
VIII can be realized.

TABLE VIII

Chemical and Physical Differentiations[a]

Direct Measurement

  Soluble Species

  NO_3^--N, NO_2^--N, NH_3-N, and soluble unoxidized
    nitrogen, $N_S(UO)$

  Total Constituents

  Total unoxidized nitrogen, $N_T(UO)$

Indirect Measurement

  Suspended organic nitrogen = $N_T(UO)-N_S(UO)$

  Soluble organic nitrogen = $N_S(UO)-(NH_3-N)$

[a]See Ref. (15)

TABLE IX

Methods of Analysis

Total Unoxidized Nitrogen, $N_T(UO)$[a]

Unfiltered sample subjected to Kjeldahl digestion.
Distillation and analysis for NH_3-N

Precision poor because of sampling difficulty.

For mean values of 382, 702 and 265 μg $N_T(UO)$/1, standard deviations of 9.02, 97.3 and 48.6 μg/1, respectively, were obtained.

Soluble Unoxidized Nitrogen, $N_S(UO)$[a]

Sample filtered through 0.45 μ membrane filter followed by Kjeldahl digestion of filtrate, distillation and analysis for NH_3-N recovered.

Precision good, for a mean value of 340 μg $N_S(UO)$/1, a standard deviation of 12.3 μg/1 was obtained.

Sensitivity as for NH_3-N

Ammonia-Nitrogen, NH_3-N

1. Distillation and Nesslerization[a]

Sensitivity - good precision down to 35 μg of NH_3-N/1 if sufficient sample was distilled.

Interferences - easily hydrolyzable amino compounds

2. Alkaline oxidation of NH_3 to NO_2^-[b]

a. Sample + NaOCl + base converts NH_3 to NO_2^-

b. NO_2^- determined colorimetrically[b]

Sensitivity - down to 1 μg NH_3-N/liter determined in this manner

TABLE IX (Continued)

Interferences - easily hydrolyzable
amino compounds

[a] See Ref. (15)
[b] See Ref. (52) and Table XI

Other methods of analysis for NH_3-N based on colorimetric methods are available and are given in Table X.

TABLE X

Ammonia Nitrogen Analyses

Pyrazolone[a]
    a. Sample + chloramine-T + pyridine pyrazolone
    b. Extract color with CCl_4
    c. Measure absorption at 450 nm

Sensitivity - down to 50 µg NH_3-N/liter

Phenol-Hypochlorite Method[b]
    a. Sample + phenol-sodium nitroprusside reagent + alkaline hypochlorite react to form an indophenol moiety.
    b. After 25 minutes read absorption at 630 nm

Sensitivity - in the range of 1 to 10 µg NH_3-N/liter

Interferences - decreased response to nitrogen in materials because of incomplete hydrolysis

[a] See Ref. (53)
[b] See Ref. (54)

TABLE XI

Analysis for Nitrite Nitrogen[a,b]

Diazotization and Coupling

1. Sample + sulfanilic[b] acid in acid solution to form diazotized compound

2. Couple diazotized compound with naphthylamine[b]

3. Read absorbance at 520 (540)[b] nm

Range - 2-250 µg NO_2^--N/liter (1 µg/liter)[b]

Interferences - free available chlorine and NCl_3, also Fe^{3+}, Hg_2^{2+}, Ag^+, Bi, As^{3+}, Pb, Au, chloroplatinate and meta vanadate

[a] See Ref. (55)

[b] See Ref. (52) The use of sulfanilamide and N-(1-naphthyl)-ethylenediamine, respectively are claimed to improve sensitivity and reliability.

A very sensitive method for nitrite is available based on diazotization and coupling which leads to a colored substance that absorbs between 520 and 540 nm. Many interferences are present in this test, as indicated in Table XI (52, 55). This procedure may also be used for nitrate (see Table XII).

Several nitrate (see Table XII) procedures are available which are based on colorimetric reactions (56, 57). However, a technique which allows the conversion of NO_3^- to NO_2^- by passing through a cadmium amalgam reductor column, avails the analyst of a sensitive method (see Table XII and Refs. 58, 59). A number of interferences are avoided by this procedure, e.g., hydrolyzable organic nitrogen compounds, nitrites, sulfides and others (58).

TABLE XII

Analysis for Nitrate Nitrogen

Chromotropic Acid Method[a]

1. Sample + sulfite-urea + antimony sulfate + chromotropic acid - H_2SO_4 reagent
2. In 45 minutes color develops and absorption read at 410 nm

Sensitivity - in range of 0.1 to 11 μg $NO_3^--N/1$

Interferences - none from Cl^-, Fe^{3+} and a host of anions and cations

Brucine Method for Nitrate[b]

1. Add sample to H_2SO_4 (1:4)

2. Add brucine-sulfanilic acid reagent and heat for 20 minutes

3. Read absorption at 410 nm, color stable

Sensitivity - 0.05 to 0.8 mg $NO_3^--N/liter/$ limit 0.01 mg/liter

Interferences - possibly organic nitrogen compounds if hydrolyzable

Reduction of NO_3^- to NO_2^-[c]

1. Pass sample through a Cd-amalgam column converting NO_3^- to NO_2^-

2. Analysis as for NO_2^--N via diazotization and coupling using sulphanilamide and naphthylethyldiamine[d]

Sensitivity - down to 1.5 μg $NO_3^--N/liter$

Interferences - none for NO_2^-, amino acids, urea, $S^=$

[a]See Ref. (56)

[b]See Ref. (57)

[c]See Ref. (58). Also see Ref. (59) for a similar technique

[d]See Table XI

A number of automated procedures may be carried out in determining nitrogenous materials in the environment (see Table XIII). The sample may be converted to ammonia by reductive pyrolysis, followed by a micro coulometric titration (60). It is in fact an automated method which may possibly be replacing the standard Kjeldahl method. The Auto-Analyzer has been used to determine ammonia nitrogen by means of the phenolhypochlorite method (54). The specific ion electrode for nitrate is also available.

TABLE XIII

Automated Methods for Nitrogenous Compounds

Microcoulometric Titration[a] - Total Nitrogen

1. Converts all N forms to NH_3 by reductive pyrolysis

2. Microcoulometric titration with electro-chemically generated H^+

3. Responds to amines, nitrates, proteins, nitro compounds and ureas

4. Considered a replacement for the Kjeldahl Method

Sensitivity - down to 200 µg N/liter & 10 µg/l potential lower limit

AutoAnalyzer

Most colorimetric techniques can be adapted. NH_3-N[b] via phenol-hypochlorite method is in use.

Specific Ion Electrode for NO_3^--N

Down to 140 µg NO_3^--N/liter detectable

[a]See Ref. (60)
[b]See Ref. (54)

An evaluation of the methods of analysis for inorganic nitrogen shows that adequate sensitivity is available; however, distinction between ammonia and easily hydrolyzable amino compounds still presents a problem. Adequate separation of various physical forms is still wanting and presents similar problems as discussed above for phosphorus analysis.

The advent of modern analytical instrumentation and separation techniques offers an opportunity to separate organic nitrogen and phosphorus substances which otherwise would not be resolvable. Two gas chromatographic procedures illustrate these developments. Ribonucleotides are converted to tri-methylsilyl derivatives which are separated on a methylsilicone fluid (DC430) column (61). Separation of amino acids by conversion to their respective methylthiohydantoin derivatives has been reported by Attrill et al (62). Twenty-two amino acids can be separated by this method. The recent development of liquid chromatography should be a potentially helpful technique for the resolution of water soluble organic nitrogenous and phosphorus-containing substances. For polymeric materials, molecular exclusion (gel permeation) chromatography is the method of choice. No doubt the future successes in this area will be attributable to these techniques.

CONCLUSIONS

Many techniques are available for the analysis of phosphorus and nitrogenous compounds. However, the specificity is not satisfactory in a large number of situations. A combination of modern analytical and separation techniques will be needed to determine the precise species present in environmental samples. The possibility of perturbation of the sample is great. Although no discussion of this subject was undertaken here, it is pointed out that much effort has been expended in adequate sample collection, storage and chemical and physical preparation prior to analysis.

The correspondence between the results of the analytical techniques and their meaning with respect to environmental phenomena was discussed above. However, this obliges the investigator to be primarily concerned with environmental effects and to tailor the methods toward this end. The role of nutrients has caused some controversy recently because there is a lack of agreement on the causes of eutrophication. Recently, Kuentzel (63) reviewed this subject. Thus, the role of analysis and identification, along with creative evaluation of the results will be needed in order to shed light on the function of the biogeochemical cycles for nitrogen, phosphorus carbon and other vital elements. This knowledge will hopefully lead to a better understanding of eutrophication and ultimately the intelligent management of man's impact on the environment.

LITERATURE CITED

1. Eutrophication: Causes, Consequences, Correctives, National Academy of Sciences, Washington, D. C., 1969, pp. 3-13.

2. W. Stumm and J. J. Morgan, Aquatic Chemistry, Wiley-Interscience, N. Y., 1970, pp. 515,516.

3. A. L. Lehninger, Biochemistry, Worth Publishers, N. Y., 1970, chaps. 10 and 14.

4. G. E. Hutchinson, A Treatise on Limnology, V. 1, J. Wiley, N. Y., chap. 12.

5. F. R. Hayes and J. E. Phillips, Limnol. Oceanogr., 3, 459 (1958).

6. W. Abbott, Proc. 4th Int. Conf. Water Pollution Res., Pergamon, N. Y., 1969, pp. 720-39.

7. D. E. Armstrong, D. E. Spyridakis and G. F. Lee, Preprints of Papers Presented at 157th National Meeting, American Chemical Society, Div. of Water, Air and Waste Chemistry, April 1969, p. 157.

8. J. E. Phillips, in Principles and Applications
    in Aquatic Microbiology, (H. Heukelekian and
    N. C. Dondero, eds.), Wiley, N. Y., 1964,
    pp. 61-81.

9. D. J. Cosgrove, in Soil Biochemistry, (A.D.
    McLaren and G. H. Peterson, eds.), Marcel
    Dekker, N. Y., 1967, pp. 216-228.

10. S. Olsen, in Chemical Environment in the
    Aquatic Habitat, (H. L. Golterman and R. S.
    Clymo, eds.), N. V. Noord-Hollandsche
    Uitgevers Maatschappij, Amsterdam, 1967,
    pp. 63-105.

11. W. Ohle, Z. Angew. Chem., 51, 906 (1938).

12. J. D. H. Williams, A. L. W. Kemp and A.
    Mudrochova, Preprints of Papers Presented at
    161st National Meeting, American Chemical
    Society, Div. of Water, Air and Waste
    Chemistry, March-April, 1971, pp. 48-52.

13. W. T. Lammers, in Water and Water Pollution
    Handbook, V. 2, (L. L. Ciaccio, ed.), Marcel
    Dekker, N. Y., 1971, chap. 12.

14. E.R. Purvis and W. J. Hanna, Soil Sci., 67,
    47 (1949).

15. D. Jenkins, J. Water Pollution Control
    Federation, 39, 159 (1967).

16. D. A. Wentz and G. F. Lee, Environ. Sci.
    Technol., 3, 750 (1969).

17. S. R. Olsen and L. A. Dean, in Methods of
    Soil Analysis, Part 2 (C. A. Black, ed.) Amer.
    Soc. Agron., Madison, Wis., 1965, pp. 1035-49.

18. D. W. Menzel and N. Corwin, Limnol. Oceanogr.,
    10, 28 (1965).

19. American Soap and Glycerine Producers
    Committee, J. Am. Water Works Assoc., 50,
    1563 (1958).

20. J. D. H. Strickland and T. R. Parsons, A Manual of Sea Water Analysis, Bull. No. 125, Fisheries Research Board of Canada, Ottawa, 1960, pp. 41-46.

21. L. J. Greenfield and F. A. Kabler, Bull. Mar. Sci. Gulf Carib., 4, 323 (1954).

22. H. W. Harvey, J. Marine Biol. Assoc. United Kingdom, 27, 337 (1948).

23. J. Murphy and J. P. Riley, Anal. Chim. Acta., 27, 31 (1962).

24. M. Jean, Compt. Rend., 240, 2237 (1955); Anal. Chim. Acta., 14, 172 (1956) and 31, 24 (1964).

25. D. L. Johnson, Environ. Sci. Technol., 5, 411 (1971).

26. J. D. H. Strickland and T. R. Parsons, Bull. No. 167, Fisheries Research Board of Canada, Queens Printer, Ottawa, 1968.

27. J. C. vanSchouwenburg and I. Walinga, Anal. Chim. Acta., 37, 271 (1967).

28. M. L. Jackson, Soil Chemical Analysis, Prentice-Hall, Englewood Cliffs, N. J., 1958, pp. 134-82.

29. Standard Methods for the Examination of Water and Wastewater, 13th ed., Amer. Public Health Assoc., Washington, D. C., 1971, pp. 518-534.

30. K. Sugawara and S. Kanamori, Bull. Chem. Soc. Japan, 34, 258 (1961).

31. K. Sugawara and S. Kanamori, Bull. Chem. Soc. Japan, 37, 1358 (1964).

32. Chong Hun Won, Nippon Kagaku Zasshi, 85, 859 (1964).

33. D. P. Lundgren, Anal. Chem., 32, 824 (1960)

34. Technicon AutoAnalyzer Bibliography,1957-67;
    Bibliography on Air and Water Quality, Pub. No.
    842-R-1-9-5W, Technicon Corp., Tarrytown, N.Y.

35. W. C. Butts, Anal. Lett., 3, 29 (1970).

36. R. F. Christman and R. A. Minear, in Organic
    Compounds in Aquatic Environments, (S. D. Faust
    and J. V. Hunter, eds.), Marcel Dekker, Inc.,
    New York, 1971, Chap. 6.

37. W. R. Sistrom, Microbial Life, Holt, Rinehart
    and Winston, Inc., New York, 1962, p. 68.

38. G. K. Reid, Ecology of Inland Waters and
    Estuaries, Reinhold Publishing Corp., N. Y.,
    1961, pp. 183-187.

39. W. H. Durum, in Water and Water Pollution
    Handbook, V.1, (L. L. Ciaccio, ed.), Marcel
    Dekker, Inc., New York, 1971, pp. 2-6.

40. Cleaning Our Environment, American Chemical
    Society, Washington, D. C., 1969,pp. 109-120.

41. Ref. (39) p. 16.

42. G. F. Hutchinson, A Treatise on Limnology, V. 1,
    Wiley, N. Y., 1957, p. 853.

43. H. A. Painter, in Water and Water Pollution
    Handbook, V. 1, (L. L. Ciaccio, ed.), Marcel
    Dekker, Inc., New York, 1971, Chap. 7.

44. G. F. Lee, Eutrophication-Occasional Paper No. 2,
    PB 197 697 (Clearing House), Water Resources
    Center, Univ. of Wisconsin, Madison, Sept. 1970.

45. D. H. Kohl, Georgia B. Shearer and B. Commoner,
    Science, 174, 1331 (1971).

46. P.H.S. Drinking Water Standards, U. S. Dept. HEW,
    Washington, D. C., 1962.

47. J. D. Rhoades and L. Bernstein, in Water and
     Water Pollution Handbook, V. 1, (. L. L. Ciaccio,
     ed.), Marcel Dekker, Inc., New York, 1971,
     Chap. 3.

48. D. H. K. Lee, Nitrates, Nitrites and Methemo-
     globinemia, Envir. Review No. 2, PB 192779
     (Clearing House) National Inst. of Health, Dept.
     HEW, Research Triangle Park, North Carolina,
     May 1970.

49. G. D. McKee, L. P. Parrish, C. R. Hirth, K. M.
     Mackenthun and L. E. Keup, Preprints of Papers
     Presented at 158th National Meeting, American
     Chemical Society, Div. of Water, Air and Waste
     Chemistry, New York, Sept. 1969, pp. 149-154.

50. G. F. Lee, Preprints of Papers Presented at
     158th National Meeting, American Chemical
     Society, Div. of Water, Air and Waste Chemistry,
     New York, Sept. 1969, pp. 143-148.

51. G. E. Hutchinson, A Treatise on Limnology, V. 1,
     Wiley, N. Y., 1957, p. 891.

52. F. A. Richards and R. A. Kletsch, in Recent
     Research in the Fields of Hydrosphere, Atmosphere
     and Nuclear Geochemistry, Maruzen Co., Tokyo,
     1964, pp. 65-81 (cf CA, 64, 17249g).

53. J. M. Kruse and M. G. Mellon, Anal. Chem., 25,
     1188 (1953).

54. J. E. Harwood and A. L. Kühn, Water Research, 4,
     805 (1970).

55. Ref. (29) pp. 240-243.

56. P. W. West and T. P. Ramachandran, Anal. Chim.
     Acta, 35, 317 (1966).

57. D. Jenkins and L. L. Medsker, Anal. Chem., 36,
     610 (1964).

58. A. W. Morris and J. P. Riley, Anal. Chim. Acta,
     29, 272 (1963).

59. R. J. Elliot and A. G. Porter, Analyst, 96, 522 (1971).

60. R. T. Moore and J. A. McNulty, Environ. Sci. Technol., 3, 741 (1969).

61. T. Hashizume and Y. Saski, Anal. Biochem., 15, 199 (1966).

62. J. E. Attrill, W. C. Butts, W. J. Rainey, Jr., Anal. Lett., 3, 59 (1970).

63. L. E. Kuentzel, Envir. Lett., 2, 101 (1971).

Current Topics in Pharmaceutical Analysis

Edited by **Edward M. Cohen**

ANALYTICAL REQUIREMENTS OF AUTOMATED
PHARMACEUTICAL ANALYSIS

By The Late
Andres Ferrari
Damon Corporation
Needham Heights, Massachusetts

EDITOR'S NOTE: Because of the tragic and untimely
death of Mr. Ferrari, the complete manuscript of his
paper is not available. The following is the abstract
as printed in the Program.

The automation of pharmaceutical analysis was
seriously undertaken in 1957 when a system and methodology for the automated analysis of streptomycin and
penicillin were presented at the 132nd Annual Meeting
of the American Chemical Society in New York City.
Since then a wealth of literature in the form of individual papers and monographs of conferences at the New
York Academy of Sciences and elsewhere have testified
both to the need for and concern of the scientific
community to provide better analytical procedures and
means to the control of and manufacture of pharmaceutical products. In the past fourteen years, changes
for the better in the analysis of pharmaceutical products have been enacted by the U. S. Pharmacopoeia,
i.e., single tablet assays. In addition, the pharmaceutical drug manufacturers have developed and improved
upon analytical procedures and adapted many of these
for increased productivity and quality control in their
laboratories. Several papers in the past have indicated that these newer concepts can provide information
on drugs and their effects upon living matters, as well
as to elicit information on the mechanisms by which

they exert their **action** in the animal body and on the normal metabolic processes of microorganisms and parasites.   The above will be discussed and reviewed.

A Look Ahead Toward Possible Future Analytical
Requirements by FDA

Daniel Banes

Director, Office of Pharmaceutical Research
and Testing, Bureau of Drugs
Food and Drug Administration

Even the most venturesome clairvoyants, with their
heads in the clouds and their gaze fixed on futurity, must
stand firmly on the ground of the present; and therefore
they generally open their dark parable on the shape of
things to come with a recitation depicting the status of
things as they are, as accurately as circumstances permit.
In that tradition, I take as my point of departure a brief
exposition on the current analytical requirements of FDA
before embarking on my personal prognostication fore-
shadowing the lineaments toward which future requirements
might evolve. Please note at the outset my emphasis on the
first person singular, and the disclaimer such usage implies.
The prophecies I offer on this program are my own, and do
not necessarily adumbrate the developing attitudes and
actions of the sponsor.

It seems to me that the purpose, need and desirability
of establishing valid analytical criteria for examining
drug products are crystal clear. By decree of Congress, as
expressed in the Federal Food, Drug and Cosmetic Act, all
adulterated and misbranded drugs are prohibited from interstate commerce. By delegation of responsibility in the
Executive Branch, the Food and Drug Administration is the
governmental body authorized to enforce that law. FDA is
a scientific regulatory agency, and it relies upon verified
analytical procedures to differentiate between drugs that
are properly constituted and those that are adulterated;
between drugs that are what they purport to be and those

that are misbranded.

The nature of the tests and methods of assay to be
utilized in this pursuit are set forth most explicitly in
Section 507 of the Federal Food, Drug and Cosmetic Act,
which deals with the certification of antibiotics. First,
the tests and methods of assay must be adequate to determine
compliance with appropriate standards of identity, strength,
quality and purity. Secondly, they should provide for ac-
ceptance or rejection of a batch within the shortest time
consistent with the purposes of the law. And thirdly, both
the standards of identity, strength, quality and purity and
the tests and methods of assay must be published in detail
as Federal regulations, so that all interested parties may
know immediately the full range of FDA's analytical requirements for each certified antibiotic preparation.

I linger over the language of these certification provisions because their significance for the future may even
far transcend their present importance. It has been proposed in several influential quarters that all drugs, not
only those containing insulin and antibiotics, be made
subject to certification. If Congress were to enact such
legislation, the dicta in Section 507 may well become the
models for analytical requirements governing the standardi-
zation of all drug products in interstate commerce. It
will be worthwhile, therefore, to enlarge upon the implications of Section 507, and to examine real-life analytical
operations under its provisions in both the National Center
for Antibiotic Analysis and the industry laboratories.

Before any certifiable drug may be distributed legally,
its sponsor must demonstrate its safety and efficacy. He
must further provide data on the manufacturing process as-
suring that the product will be uniform from batch to batch,
as well as specifications for all raw materials used, and a
description of adequate tests and methods of assay to
determine compliance with proposed standards of identity,
strength, quality and purity. Scientists in NCAA review
the manufacturing data and the proposed specifications and
standards to determine whether they are appropriate.
Furthermore, they empirically check the tests and assay
methods by applying them to real samples of the intended
product. If NCAA agrees that all is in order, the standards
and specifications, the tests and assay will be incorporated
in a monograph and published in regulations providing for

certification of the product. However, if NCAA takes issue
with the manufacturer's analytical proposals, no regulation
can be written until the objections are satisfactorily
resolved. On the basis of their experimentation and previous experience, NCAA may suggest modifications or alternative procedures. The manufacturer may then submit new
counterproposals, which are again scrutinized critically
in NCAA. This dialogue is rarely protracted. Eventually
it evolves use-tested monographs and regulations which are
acceptable to the manufacturer as a set of realistic
criteria for quality control and are FDA's yardstick for
batch certification.

Among the important considerations bearing upon the
integrity, safety and efficacy of an antibiotic drug is the
stability of the product. The antibiotic regulations require that each certified preparation bear an expiration
date based upon continual testing of the drug over a prolonged period under normal conditions of storage. The
analytical methods employed to demonstrate stability or
lack of it must be specific enough to differentiate between
the intact drug substance and decomposition moieties, and
sensitive enough to detect the degradation compounds as
they are generated. Here, again, NCAA conducts studies
parallel to those in the producer's laboratory.

Since every dosage form of every certifiable product
is subjected to such double-check processes, no situation
can arise wherein an inappropriate additive interferes with
the prescribed method; such a state of affairs would render
that particular batch ineligible for certification, and the
manufacturer would be made aware of his contretemps by his
own findings, or by NCAA.

Furthermore, an assay method proved suitable for a
particular dosage form is never thoughtlessly accepted as
applicable to other dosage forms, or even to a modification of the same dosage form. Every proposed method is
tested for accuracy and practicability by applying it to
the individual product in question.

If subsequent laboratory experience shows that the
regulation promulgated does not assure a product which is
safe, efficacious, and uniform, the Food and Drug
Administration may suspend certification. It may also
take action to modify the regulation by means of appropriate

amendments published in the Federal Register, if the manufacturer can demonstrate that these changes will provide the
required assurances. Prompt publication of new regulations
and amendments continuously provides current official
standards for all of the antibiotic drugs eligible for
certification and distribution.

The Antibiotic Regulations require that the manufacturer test all batches to be certified and submit his
data to NCAA prior to distribution. He may utilize procedures other than those in the official monographs, but
only if he has demonstrated experimentally that his alternative methods yield equivalent results. In any case of
conflicting data, those obtained by the official tests and
assay prevail. Thus, the certification requirements demand
that the manufacturer will have used acceptable raw materials
in fabricating his products, will have ascertained that every
batch complies with official specification and standards,
and will have demonstrated the stability of the article
throughout its expiration period. Now, if certification is
to be extended to non-antibiotic preparations as well,
presumably these stringent pre-marketing requirements will
then be mandatory for all drug products.

In theory, such a development should not materially
affect the analytical burden placed upon the drug industry
laboratories. For all of the research and control operations
mentioned -- careful examination of raw materials for acceptability, studies on product stability to establish appropriate expiration dating, and use of scientifically sound
specifications, standards and test procedures to ensure the
quality and integrity of finished products - - all of these
activities are already general requirements in the FDA
Regulations on Current Good Manufacturing Practices. However, there is reason to believe that observance of CGMP
principles is presently less than universal, except where
certifiable drugs are involved. In the latter case,
failure to comply would delay distribution and sale of the
product, and economic interest therefore dictates prompt
completion of the appointed tasks. Thus, it can be predicted with confidence that total certification would
result in heightened analytical endeavors in both FDA and
industry laboratories. Accelerated utilization of automated equipment and computerized systems would inevitably
follow.

An increase in the use of automation and computerization because of more demanding FDA analytical requirements
is a foregone conclusion, even if a total certification
program never eventuates. The recently adopted content
uniformity requirement practically guarantees prompt veri-
fication of that forecast. The dimensions and perplexities
of the problems associated with lack of individual tablet
content uniformity were first recognized by the FDA
laboratories. It was noted that assays performed on a
certain batch of cortisone acetate tablets yielded un-
accountably discordant results. The product was found to
contain an excess of the steroid when examined by the official
U.S.P. assay. But repeated check analyses on different 20-
unit groups of tablets from the same batch by the same method
afforded values ranging from subpotent through satisfactory
to superpotent. These erratic data could not be correlated
with the weights of the tablets, which were consistent
throughout the batch. However, eventual analysis of the
individual tablets revealed a drastic variability in their
cortisone acetate content.

The disclosures on the cortisone acetate tablets were
buttressed by similar FDA findings in supplementary studies
on diverse pharmaceutical products. On the basis of the
accumulated date, FDA concluded that acceptable assay
values and satisfactory weight variation data are not in
themselves warranties of drug quality. If the effective
drug substance is poorly distributed through the granulation, it is possible for many of the tablets punched from
the mass to fall far outside the permissible potency limits,
even though the weights of the individual tablets are
practically identical and the official assay shows that the
average value for twenty or more tablets is acceptable.

For many classes of pharmaceuticals, this state of
affairs might be regrettable, but not of serious concern.
However, the dosage regimen of extremely potent drugs like
the anticoagulants and the cardioactive glycosides, whose
therapeutic dose is not much less than a toxic dose, must
be kept within strictly defined limits. For such medica-
ments, wide tablet-to-tablet fluctuations in the quantity
of the active substance conceivably could be catastrophic.

Such alarming speculations are not merely theoretical
conjecture. In a current monitoring program on digoxin
tablets, the FDA National Center for Drug Analysis (NCDA)

has reported that 47% of the batches investigated did not
comply with the requirements of the U.S.P. monograph,
chiefly because of failure in the content uniformity test.
In one sample, NCDA found tablets containing twice the
declared quantity of digoxin, together with a large number
of tablets containing 60-70% of the declared quantity.

Such difficulties are not restricted to digoxin
tablets. Surveillance of digitoxin preparations and of
ethinyl estradiol tablets has uncovered similar problems
with these articles. The most erratic lot encountered has
been a batch labeled to contain 0.05 mg ethinyl estradiol
per tablet. Ninety tablets were analyzed individually.
The average assay for the ninety tablets was 99.4% of the
declared quantity of the active ingredient. But the range
of assays was 37-252%; 10 of the 90 tablets contained less
than 50% of the declared quantity; and 15 contained more
than 150%. It is readily apparent that the content uniformity test is a necessary analytical criterion for the
quality control of high potency drugs.

In establishing the National Center for Drug Analysis,
FDA has significantly expanded its capacity for gauging
the quality of drug control throughout the industry by
simultaneous comparison of similar products. Because of
the volume and variety of its work and the vital implications of its findings to regulatory decisions, NCDA has
been compelled to study the performance of many official
analytical methods in meticulous detail and to demonstrate
that its automated procedures yield data of equal validity.
Indeed, in some instances, the automated methods are
superior in trustworthiness. NCAA had arrived at the same
conclusion earlier in comparing its automated methods for
penicillin preparations with manual iodometric assays, and
several other laboratories have reported similar findings.
Automated systems for drug analysis can be expected to
proliferate mightily.

Another vast and largely uncharted problem area lies
in the domain of drug bioavailability. To determine the
concentration of a drug like digoxin in the circulating
blood of subjects after a dose of 250 micrograms is a
formidable task. In the past, chromatographic procedures
have been utilized to determine microgram quantities of
drug substances in body fluids and tissues. Even more
sensitive methods, capable of measuring nanogram and

picogram quantities will be needed. Satisfactory radio-
immune assays have been developed for the determination
of digoxin in bioavailability studies, and electron spin
resonance has been proposed as a feasible technique for
quantitating alkaloids and related bases at the desired
level of sensitivity.

It can be anticipated that FDA will continually
require more extensive data on the concentrations of active
principles and metabolites in body fluids after administration of new and potent drugs, to ensure that the levels
observed are comfortably below concentrations associated
with toxic reactions. These reasonable, but monumental
requirements will create a need for novel ultra-micro
analytical methods. Given the present state of the art in
analytical chemistry, and the superlative resourcefulness
and ingenuity it reflects, I am sure that chemists of the
future will respond to the challenge with brilliant
success.

DATA AND INFORMATION HANDLING SYSTEMS IN THE PHARMACEUTICAL, ANALYTICAL, AND QUALITY CONTROL ENVIRONMENTS

H. Stelmach

Analytical Research Department
ABBOTT LABORATORIES
North Chicago, Illinois

The present upsurge in the use of digital computers is significantly changing the operational procedures in scientific laboratories and nowhere is this more prevalent than in the analytical research and quality control laboratories. Where valuable man-hours were being wasted with routine repetitive bookkeeping and calculating, computers are now utilized to sort, store, record, and calculate.

Computers are amazingly versatile machines capable of accomplishing an infinite variety of computational, logical, and control operations. For instance, the same instrument can keep records of personnel, prepare the payroll, compute statistical parameters, collect and process data from instruments, and perform curve fitting and plotting. Moreover, the computer is a tireless machine which is content to calculate continuously and consistently, and with blinding speed (10^5-10^6 instructions per second). This marvelous versatility is something with which we are not familiar in the normal purchase of laboratory instrumentation. We plan for and purchase instruments for our analytical laboratories to accomplish one particular application. We expect an atomic absorption spectrophotometer to analyze for metals, our gas chromatograph to produce chromatograms, and the spectrophotometer to turn out spectra. We anticipate these results immediately upon installation of the instrumentation. However, the purchase of a computer, a relatively simple device, provides nothing at all. Without a specific program or instruction the computer cannot accomplish even the most

fundamental computation. Thus, although grand and glorious
applications are generally perceived to be effected with the
purchase of a computer, they rarely materialize without
adequate and proper planning.

Computer vendors frequently point to payroll applications and credit card services and tell us how the available
computer and ready-made programs will, or can with minor
adjustments, solve all of our problems. This may be the
case where the applications are general and very little
modification of ready-made programs is necessary; however,
we have found that these programs do not suit our needs;
they do not perform the required calculations for our specific needs nor is the format of print-out satisfactory.

It is essentially impossible for vendors to develop
automated systems for analytical instrumentation which will
satisfy all users. And while there may be many laboratories,
even within the same company, that utilize computers with
the same analytical instruments, each has requirements unique
to their own usage. Thus, we have found that a large gap
existed between our understanding of the theories of computer
programming and our abilities to put them into practice.

Another hard lesson we learned was that in order to
accomplish anything meaningful, we would have to develop
people with computer skills within each department or at
each organizational level. Mathematics and computer personnel rarely have so broad a background as to enable them to
deal effectively with the idiosyncrasies of both computer
planning and programming, and pharmaceutical and chemical
processes. We found it advantageous to use these people as
resources and develop scientific people with the required
skills to allow for intelligent modification of available
systems or to develop entirely new systems for our specific
applications.

Much to our dismay, in our initial endeavors, the
analytical department developed "computer fever" and attempted to computerize everything to obtain maximum utilization of this instrument. We had illusions of computer usage
revolutionizing the analytical laboratories and freeing its
scientists. This, combined with an urgency to justify our
expenditure of time and money, led us to try to utilize a
complicated and inefficient computer system to perform a
simple manual operation. The best example of this was

attempting to use the computer for data handling when we
had too small a volume of data. In this case, as in the
case where there is no common denominator in the methodology,
we found that a programmable desk calculator was better able
to satisfy our needs.

You may ask why the difficulties we experienced happened.
The main reason was we neglected to properly plan the applications to which the computer was to be put and what would
be necessary to implement these applications prior to arrival
of a time-sharing terminal in the laboratory.

You may also ask how some of the problems we had could
have been avoided. I don't have any hard-and-fast rules,
but prior to the start of projects one should ascertain that
the computer is a correct way to proceed. The following
questions may help:

1. Can a significant number of man-hours be saved by
   computerizing the operation?

2. Can an accomplishment be realized with the computer
   that would be impossible to attain otherwise?

3. Can something important be accomplished which has been
   avoided previously?

Then, after deciding that the proper approach is computerization, consider some of the criteria for a successful
computer system.

1. The first point is the relevance of the system to serve
a need. It is critical to clearly state the objective of a
system, who will use it, how will they relate to it, and how
the system will satisfy the stated objective. Sometimes one
tries to solve a problem by computerizing it, assuming that
the computer itself is the cure rather than the machine to
be used to solve the problem. These situations generally
end up as a computerized confusion rather than a manual mess.
And the computerized problem is more difficult and more
expensive to solve than it would have been if it had been
approached properly in the first place.

Another example is the executive who decides what his
employees need to function more effectively and more efficiently and then requests the programming systems staff to

define, design, and implement a system. When the system is
given over to the employees, they view it as something else
which they don't need. We probably do not have to look very
far within our respective organizations to recall some computer generated reports which were neatly filed away or thrown
away without ever being read because we considered them
irrelevant to serve our needs.

The very nature of computer planning involves people of
many disciplines; these are the user, the systems programmer,
the hardware people, and the applications programmer. These
people should be brought together at the earliest possible
time in planning process and kept in constant contact for
the duration of a project. They should work and plan together, be aware of each other's capabilities and limitations,
and most importantly, they must keep their efforts directed
towards the needs of the user.

2. A second major criterion to consider is flexibility. One
aspect of flexibility is the capability of a given system to
handle normal growth in the future. Thus a chosen system
must have maximum utilization at any point in time, but
should be capable of extension in a modular way so that equipment and other necessary elements can be added as growth occurs. A modular configuration will enable expansion without
major reprogramming and redesign and thus keep expenses
minimal.

A second aspect is the capability to handle the inevitable changes in the planning process or in the operations
of the company. Pharmaceutical manufacturers constantly
strive to improve; therefore, change will be inevitable. If
the system is not flexible to handle the changes, it quickly
becomes highly cost ineffective and obsolete.

3. This brings up the criteria of response time. Is immediate turn-around time required and can it be achieved with the
system available? If not, it may be better not to proceed
than to produce a system which is not useful within the actual
time frame.

Monitoring of a fermentation process requires immediate
response-time to preserve the integrity of the process as
does monitoring of waste effluent. It is useless to know
hours or days later the process was not functioning properly.
One cannot retrieve effluent which passed into Lake Michigan

two days earlier. On the other hand, it is not critical to
have the completed calculations from a batch of tablet assays
simultaneously with the completion of the assay because the
formulation and tableting procedure have already been completed and the process is in a pass-fail status.

4. Economy is another feature of an effective system. The
intangible results, such as increase of employee morale or
favorable responses from other parts of the company, are
difficult, if not impossible, to assign a dollar-value. On
the other hand, the tangible results can easily be measured
by
   a. Savings in time and personnel.
   b. Increased capacity to run a large number of samples
      simultaneously.
   c. Reduction in paper work.
   d. Fewer sources of human error and thus,
   e. The advantage of high accuracy and reproducibility.

5. Lastly, a successful system must operate with accuracy.
In the control laboratories, the simplest and most common
use for a computer is as a replacement for a desk calculator
in long, complex, or repetitive calculations which are required to convert physical measurements into the desired
quantities of identity, potency, purity, etc. Thus, systems
must be installed with adequate controls so that the users
can rely on the accuracy of the reports which they receive;
otherwise, the users will begin to devise manual systems to
give them the accurate data which they require. The result
of this oversight is a computerized system which is not used
and a duplicate manual system, which although it is recognized as inefficient, is nonetheless necessary in order to
enable the user to perform his function.

In summary, the criteria for a successful computer system are relevance, flexibility, adequate response time,
economy of operation, and accuracy.

Now let's look briefly at the calculator and computer
types of data handling systems and compare them.

1. Desk Calculator: With the rapid increase of technology
and manufacturer effort to make computational capability
available at minimal cost, we should first reevaluate the
desk calculator. Previously desk calculators had only the
capacity to add, subtract, divide, and multiply. Today

these instruments lie on the border of being called computers themselves. They have memory capacity and are
programmable. They can directly accept digital output of
instruments or punched tape. They have the capability to
display data digitally or graphically. In addition, many
calculators allow for modular expansion. Desk calculators
do have the disadvantages of limited computational power
and slow speed; also, there does exist the possibility of
human error.

2. Computers: Use of computers is of three general types
or any combination thereof. The first of these is the
dedicated computer or system which ranges in size from one
small computer dedicated to a single instrument to where a
single medium sized computer may service ten or more similar
instruments. The computer is activated by an instrument
control which notifies the computer to start accepting data
from a given channel at a given rate. The acquired data are
then written into a file. At the completion of the experiment, the user again activates a control notifying the
computer of completion and processing is begun. Data are
generally returned through a single teletype which
services all channels.

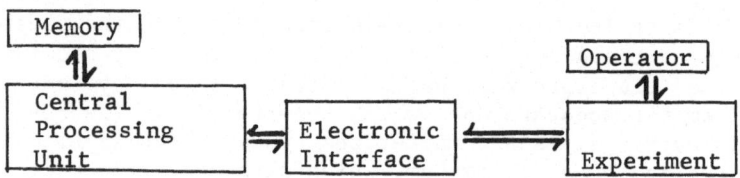

Figure 1:   Dedicated Computer Configuration

The dedicated computer approach has several inherent
features which recommend it for research applications. It
permits the operator complete freedom of action within the
computer's limitations which means that timing errors can
usually be reduced; the user does not have to be concerned
about interactions with other programs; the programs may
be written in any language compatible with the computer.
Turn around time is minimal and few, if any, scheduling
difficulties are encountered. Much of the difficulty of
these systems arises from the small memory size which
necessitates programming in machine language; thus, this

system requires much programming and interfacing effort.
The cost to obtain a larger memory or peripheral accessories
is high compared to the initial cost of the system.

The second system is the on-line computer system for
the user who requires rapid results from the system. Programs for the data processing are either resident in memory
or easily readable into memory to provide for rapid
computation. The acquired data are transmitted directly
to the computer through an electronic interface and the
resulting data transmitted directly back to the user via
teletype, plotters, printers, oscilloscopes or in any other
desired output form. In its most advanced usage, the computer may be programmed to communicate directly with the
experiment and alter the controlling electronic or electromechanical devices (any device activated by a voltage
level change).

Some advantages of this configuration are readily apparent. The real-time interaction between the experiment
and the computer allows for modification of an experiment
while it is still in progress. Secondly, there is no
human interaction with the data which leads to time delays
and possible handling errors.

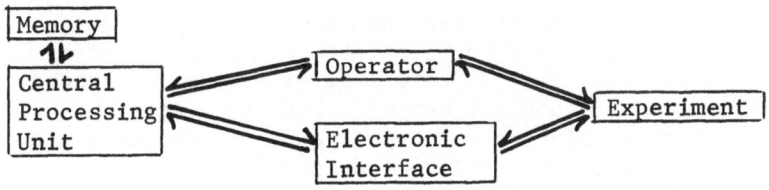

Figure 2:   On-Line Computer Configuration

A last type of computer system is the off-line computer
system which is characterized by accumulation of the experimental data through some intermediate storage such as punched
tape, magnetic tape, or punched cards. The stored data are
transmitted to the computer at some later time for data reduction and computation. This off-line data processing may
take one of two forms -- batch processing on an in-house
computer facility or the use of commercial time-sharing
services.

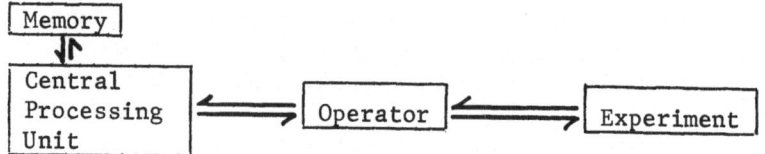

Figure 3:   Off-Line Computer Configuration

Batch processing of accumulated data may have the
following format: the data may be obtained on or transferred
to punched cards. These data cards are incorporated into a
deck of program cards; then this combined card deck is trans-
ported to the computer facility. The time required to process the program or the turn-around time may vary from a few
minutes to a few days, depending on the capacity of the
particular computer, the number of users, and the backlog
and priorities of work to be processed. Although such a
system can severely limit the versatility and dimensions of
automation, it is highly desirable as support to an on-line
control and data acquisition system.

In commercial time-sharing, data are usually accumulated
directly on paper tape and taken to a time share terminal
where access to the computer is achieved by placing a tele-
phone call to the computer. A conversational sequence between the user at the teletype keyboard and the computer
identifies the user and permits selection of the desired
system language as well as retrieval of library or stored
programs from the computer memory. The user imputs his
data and then receives a teletyped report of the analysis
within the time-span of a few minutes.

There are some distinct advantages of commercial time-
sharing computers. The commercial services are able, due to
efficient use of computer time, to maintain larger computers
that accept high-level languages. Many also provide training courses in programming and give assistance on small jobs.
In addition, a library of programs is available for general
applications in science, engineering, mathematics, etc.
These features would provide any laboratory with computer
capacity at a minimal cost. The frustrations experienced
with these systems usually center around the difficulties of
communication with an independent isolated entity.

DESK CALCULATOR

Advantages:
1. low cost
2. no loss of data
3. allows modular expansion
4. minimal maintenance

Disadvantages:
1. limited computational power
2. extremely slow
3. human error

ON-LINE COMPUTERS

Advantages:
1. real-time interaction
2. no human interface
3. available for many instruments
4. use high-level languages
5. computer control of functions

Disadvantages:
1. very high capital cost
2. possibility of transmission errors

DEDICATED COMPUTER

Advantages:
1. no scheduling difficulties
2. immediate turn-around time
3. no human interface
4. no interaction with other programs

Disadvantages:
1. high capital cost
2. small memory size
3. expensive to expand

OFF-LINE COMPUTERS

Advantages:
1. use high-level languages
2. moderate cost
3. efficient use of computer time
4. programs and training available

Disadvantages:
1. longer turn-around time
2. human interface required
3. user isolated from computer

TABLE 1: COMPARISON OF CALCULATORS AND COMPUTERS

I'd like now to turn to the work at Abbott Laboratories
and briefly describe some of our applications and experiences.
We have two independent IBM 1800 computers. Both of these
computers are available 24 hours a day for on-line data acquisition and computation. In addition, they handle batch
processing jobs for eight to nine hours daily and are also
used nights and weekends for long batch jobs and independent
batch processing. Besides the use of these computers, considerable use is made of commercial time-sharing services.

Each of the systems at Abbott Laboratories operates independently to meet separate needs. Briefly, they are (1)
Commercial time-sharing services as an introduction to computer usage and to handle less routine applications; (2) One
IBM 1800 for analytical research and control data; and (3) A
second IBM 1800 for pharmacological and clinical data. Let's
examine each of these uses separately.

Commercial Time-Sharing: The Analytical Department became aware than an increasing proportion of time was taken
up by routine calculations of gas chromatographic data and
began to consider computerized data handling. In order to
gain access to the Abbott IBM system, one should have computer programming capability within the department. Because we
did not have this capability and due to economic conditions,
the department decided to start its orientation to computers
through time-sharing and ready-made programs. Ready-made gas
chromatography computer programs were obtained with the purchase of an integrator, teletype and tape punch. The programs
provided for normalization, internal standard, and external
standard calculations. Shortly after installation of the
system, it was recognized that the programs did not handle
data with our specifications. Thus, many hours were devoted
to developing adequate programming capability and the programs
were rewritten to be consistent with our needs.

Another initial requirement was that the interaction of
the user at the computer time-share terminal be minimal, we
did not want the user to be spending twenty or more minutes
communicating with the computer prior to usage. In addition,
the systems had to be simple enough for everyone in the laboratory to be able to use them. This latter requirement was not
possible to fulfill, nor, in light of our present experience,
is it highly desirable. Just as the maintenance of an instrument is most reliably accomplished by a single individual,
a single or a few trained individuals relate to computer

systems most effectively.

Abbott's chemical development area had experienced some
of the same difficulties. They assigned one clerical worker
to write the required programs, to prepare the punched tapes
for the different assays, and to handle the interactions with
the time-sharing terminal. This method of data handling has
resulted in increased computational accuracy and has freed
the analyst for other laboratory work.

IBM 1800 for Analytical Research and Control Data: The
easiest way to describe the operation of the IBM 1800 for
on-line data acquisition from instruments is with an example.
The autoanalyzer is the most common instrument connected to
the system. Therefore, I'll take the autoanalyzer and describe what was done with this system.

Suppose, for example, there are no programs running in
the computer and no instruments are on-line. Then the computer runs in an endless circle keeping track only of time
and waiting for something to happen. When an autoanalyzer
wants service, the user in the laboratory pushes a button
sending a signal to the computer. The computer identifies
this interruption as a request for service. The user is requested to answer questions about his system. What is the
number and concentration of standards? What is the sample
rate? What is the sample dilution? After the questions are
answered, the user is requested to set the chart pen to complete the programming.

The computer turns on a "ready" light at the autoanalyzer
station, signifying it is ready to accept data. The computer
then returns to running in its endless circle. When the user
is ready to begin his run, he presses a button asking for
sampling to begin. This again interrupts the computer. It
identifies the new request and begins to look at the autoanalyzer output at the chosen time intervals. The computer
converts the analog signal to a digital value and stores this
value in a table in memory. Data collection continues until
the end of the experiment as signaled by the operator.

The choice of what to do with the data depends upon the
type of analyses being performed. Most likely, the user wants
a standard curve for the reference samples, a computation of
the concentrations of the unknowns based on the standard
curve, an index of the precision and accuracy of the analysis,

and the preparation of some kind of report. This data
handling is accomplished in one of three ways: (1) Real-
time on-line; (2) Batch processing with a resident program;
or (3) Batch processing with a program deck.

For real-time on-line data handling, the program is
resident in the computer memory and activated just previous
to the start of sampling. As the analyses progress, the
calculations are performed and the results printed out
immediately. This rapid turn-around time allows the user
to repeat samples or to make changes in the experiment while
it is still in progress. In the event of computer malfunc-
tion, it is easy to determine which data was lost and either
reassay those samples or calculate the answers from the recorder output.

A second mode for handling the data is off-line. In
this case, the data handling program is resident in computer
memory. It is not activated until after data collection has
been completed. A request is made to process the data, and
it is handled as a batch processing job. The data is printed
out at the computer facility and the user may generally pick
it up in about three hours.

The last method of handling data is through the use of
a program which is kept on punched cards. At the end of
the day, the data stored in the computer is unloaded onto
punched cards. These data cards are picked up by the user
at the computer facility, combined with the program deck,
and the entire deck is submitted for batch processing. The
turn-around time is usually one or two days, but has been
as long as four days. Therefore, the main difficulty with
this system stems from the delay in obtaining completed data.
Another difficulty arises from the need to hold the data
resident in the computer until data cards are punched. Thus,
two runs cannot be made on the same instrument in the same
day as the data from a second run would wipe out the previous
data.

As of now we have six autoanalyzers located in three
different buildings connected to this system. Presently,
we are expanding to include an additional six autoanalyzers.

A second instrument using on-line data acquisition and
computation is an automatic plate reader. This instrument
was designed and built at Abbott. It is designed to measure

the size of a zone of bacterial inhibition on a Petri dish
containing 4 or 6 evenly spaced zones. Pulses of light are
projected through the plate and counted as the plate rotates
with change in light transmission tripping the printer. At
the end of each series of plates, the data collected is
averaged and the results printed out. About 250 plates,
each containing six zones, can be read and computed in
approximately two and a half hours. Previously, it required about seven hours just to measure the diameter of
the zones.

Other on-line uses include a high resolution mass
spectrometer and instrumentation for monitoring pyrogen
testing.

The above on-line computer uses have mainly been applied where the assay procedure is highly repetitive and a
large volume of samples or data are involved. The remainder
of the analytical and quality control applications on the
IBM are performed in the off-line mode. Here the data may
be collected directly on punched tape or punched cards and
taken to the computer facility for batch processing.

These applications which have been described are in
the category of data handling, making computations, and
generating the associated reports. The high speed of operation of a computer makes it an ideal instrument for sorting
and storing information. We have taken advantage of these
capabilities in handling the data from the stability testing
program. Before describing what we can do with this system,
I'd like to outline what some of the problems are and what
we hoped to accomplish by using a computer system in this
area. The stability group is responsible for maintaining a
continuing stability monitoring program on established
products. This encompasses the testing and evaluation of
about 800 products and approximately 5000 different batches
of material which are used to make these products. It is
obvious that huge bookkeeping and reporting requirements
exist in this area. Therefore, the main intent of installing
a system was to reduce the clerical labor involved and to aid
in the retrieval of data.

After each sample is identified by its list and lot
number, the sample is assigned to a stability test schedule.
On a weekly basis, the stability test schedule is scanned and
IBM test cards are prepared for those products which require

analysis. A test card is a punched IBM card which requests
a single assay on a particular product and accompanies the
sample to the control laboratory. The analyst records his
result on the card. Then after an initial review of all
the results, all the data are filed on magnetic tape. The
stored data may be retrieved on the basis of the list and
lot number sequence. This print-out allows one to follow
the stability of a particular ingredient or product. Re-
questing information on the basis of a test code permits
one to evaluate a specific method utilized in testing.
Finally, the system can print out a monthly backlog report.

At its present stage of development, this system still
requires a high degree of human involvement with the associated possibility of error. Also retrieval is not yet as
specific as desired -- one cannot request the analysis results for one component of a multicomponent formulation without receiving the results for all other ingredients in the
formulation. Efforts are under way to improve performance
in both of those areas. And although not yet complete, I
can't imagine going back to our old method of data retrieval.
For if we did, clerical cost would rise considerably higher
than for the operation of the present system.

IBM 1800 for Pharmacological and Clinical Data: The
second IBM at Abbott has no analytical instrumentation on-
line. Most of the analyses of clinical data are handled by
outside testing laboratories. Therefore, for clinical data,
this system is used for its storing and sorting capabilities.

Four pharmacological laboratories monitor animal
studies in the direct on-line data collection mode. These
are:

1. Drug Abuse Laboratory. This laboratory is concerned
    with the determination of possible ways to abuse drugs.
    They also conduct preference screening of new drugs
    candidates against known abuse drugs.

2. Neuropharmacology Laboratory. Here the brain output of
    animals is monitored under normal and drugged conditions.
    These expermients require full computer capacity for the
    duration of the experiment, usually 16 hours.

3. Animal Work Laboratory. The ability of animals to solve
    problems is observed under normal and dosed conditions.

Parameters that can be determined are the error rate,
the number of correct solutions, the types of problems
an animal will do, how hard does the animal want to work.
This laboratory operates on a 24 hour basis.

4. Cardiovascular Laboratory. Here the polygraph signals
    are monitored by the computer and the accumulated data
    is processed immediately with a print-out of the data
    in the laboratory. Prior to the use of the computer
    in this laboratory, only two experiments were performed
    each week. Two days were spent running the test and
    the following three days were required to perform the
    necessary calculations. Presently we can complete 20
    experiments per week.

In closing, I would like to point out that these are but a
few of the data and information handling systems we have in
operation at Abbott Laboratories. And although we did experience some difficulties in putting these systems into
operation, we feel that the results were well worth the
effort spent and that further potential exists for computer
usage in the pharmaceutical industry.

References:

General or Review

C.W. Childs, P.S. Hallman and D.D. Perrin, "Application of
Digital Computers in Analytical Chemistry", Talents 16,629
(1969).

B.J. Cohen, Cost-Effective Information Systems, American
Management Assn., Inc. (1971)

J.W. Frazer, "Digital Control Computers", Anal. Chem. 40 (8),
26A (1968).

N.R. Kuzel, H.E. Roudebush and C.E. Stevenson, "Automated
Techniques in Pharmaceutical Analysis", J. Pharm. Sci. 58,
381 (1969).

J.S. Mattson, Ed., Applications of Computers to Chemical
Instrumentation, Vol. 2, Marcel Dekker.

J.H. McRainey, "Source Data Information", Automation 14 (10),
87 (1967).

S.P. Perone, "Computer Applications in the Chemical Laboratory - A Survey", Anal. Chem 43, 1288 (1971).

S.P. Perone, Digital Computers - Sizes, Shapes, Etc., Datamation March 1969.

R.J. Spinrad, "Automation in the Laboratory", Science 158, 55 (1967).

T.J. Williams, Ind. Eng. Chem. 59 (12), 53 (1967).

"Computers in Medicine and Biology", Ann. N.Y. Acad. Sci. 115 (2), 543 (1964).

"Computers and Automation Annual Buyers Guide" Comput. Automat. 17 (6), 8 (1968).

Specific
M. Blanco, S. Eriksen and M. Boghosian, "Computer Control of Drug Stability", Drug Cosmet. Ind. 1, 44 (1970).

M. Kaplan, "Computer System Design for Stability Testing Programs", Bull. Parenteral Drug Assn. 24, 23 (1970).

W.C. Kenyon, "Storage and Retrieval of Instrumental Analytical Data", Anal. Chem. 35 (12), 27A (1963).

I.S. Levine and W.G. Sutton, "Total Lab Automation", Oil and Gas J., 63, March 31, 1969.

C.H. Newman, "Computerized Stability System for Pharmaceuticals", presented at the meeting of the Parenteral Drug Assn. in New York City on February 2, 1968.

A.N. Stevens, "A Punched Card System for an Analytical Laboratory", J. Chem. Doc. 1, 74 (1961).

F.E. Tytle, "Computerized Searching of Inverted Files", Anal. Chem. 42, 355 (1970).

"Computer Automation of Analytical Gas Chromatography", Symposium held in New York City, September 8-12, 1969 published in J. Chrom. Sci. 7, 709-744 (1969).

SEMANTICS IN SPECIFICATIONS FOR DRUGS

Lester Chafetz

Pharmaceutical Research and Development Labs.
Warner-Lambert Research Institute
Morris Plains, New Jersey 07950

Several years ago I had a cousin who worked at the
William Alanson White Institute for Psychiatry, Psychoanal-
ysis and Psychology here in New York. I met several of her
friends from work on occasional visits here. When they
asked what I did for a living, I replied that I was an
analyst. This made me one of the gang, and sometimes I
could carry through an entire evening without it being discovered that I was *our* kind of analyst. Everyone understood
what "analyst" meant, but their extensional meaning differed
considerably from mine. Each of us has a personal interpretation of words. To communicate we have to arrive at a
common understanding of these terms, and failure to do this
has led to enormous confusion in the area of drug specifications, our topic here. Usually, the precise meaning of a
word becomes clear from its context in a lecture situation,
where one talks and another listens. Our problems with
semantics are most acute in the give-and-take of committee
work. These should be recognized by anyone who participates
in committees in professional societies, with the official
compendia, within a company, or in negotiation with the Food
and Drug Administration or other national regulatory agencies.
Take, for example, words like "identity", "strength",
"quality" and "purity", beloved by the FDA regulations
writers. These defy precise definition, and, certainly,
there was no intention that they be precisely defined, for
definitions change with advances in technology. Our personal
definitions, too, change with the framework used; "purity"
means one thing to us with a drug substance and another with

a dosage form.

Asking someone at a committee meeting to define his
terms is a disruptive tactic. As my panelist colleagues
can attest, I have done this on a number of occasions, with
devastating effect. In the typic *ad hoc* committee, everyone
starts out with a personal conception of what the problem
is and how it should be worked out. After a lengthy shake-
down period, involving a good deal of often eloquent and
usually pointless oratory, the majority of the *working*
members evolve a common understanding of the problem and the
meanings of the words they use in describing solutions to
it. Ultimately, a final report is issued, and the committee
is discharged with the thanks of the sponsoring body and the
relief of all concerned. The real mischief occurs when it
comes time to translating the committee report into some
kind of positive action, for only the committeemen involved
appreciate what they were doing, and they hardly ever get
around to defining the problem and the terms used explicitly.
I can think of any number of examples to illustrate this,
but, since we are limited in time, let me use as a case
history the U.S.P. *Related foreign steroids* test. This
started out - by my personal definition - as a paper chromatographic test which limited interferences in the blue
tetrazolium assay for corticosteroids, that is, it was a
means to achieve assay selectivity. Because there were
problems with its reproducibility, U.S.P. asked the P.M.A.
Quality Control Section to study the test with a view to
improving its mechanics in the light of improved technology.
After 3 years, the committee came up with a quantitative
TLC assay for corticosteroids. During this time, they never
defined *Related foreign steroids* nor indicated what should
be done with the results of their labors. Different people
in U.S.P. and N.F. interpreted their report in various ways
according to their own understandings, and the steroid
specifications in the compendia are a mess as a result.
Since official monographs serve as models for specifications
writers in the drug industry and for FDA examiners who pass
on N.D.A.'s, the confusion generated in this area -for
want of definitions- has been enormous.

It is relatively easy to do laboratory work. Through
automation and computerization, our technology affords us
the means to produce large volumes of data and dress it up
in a statistical overwrap. This productivity is two-sided;
instrumentation is no respecter of persons. While it may

enable the good scientist to increase his productivity, it
also allows the poor scientist to provide a flood of data
from ill-conceived experiments. This presentation is a
plea for more attention to be paid to the "why's" of our
projects before concentrating on the "how's." The toughest
work any of us do is thinking on company time. If there
were more of it, we wouldn't have to rework and patch tests
and standards again and again.

WHAT PROBLEMS DOES ONE INTRODUCE WHEN DIFFERENT CRITERIA
FOR ACCEPTABILITY OF UNFINISHED ACTIVES VS. THE FORMULATED
ACTIVES ARE UTILIZED?

G. J. Papariello

Wyeth Laboratories, Inc.

The particular topic that I am going to cover is
entitled, "What problems does one introduce when different
criteria for acceptability of unfinished actives vs. the
formulated actives are utilized". That is, we are aware
that when one uses a different test or analytical procedure
for testing and releasing an active component as a raw
material as compared to testing and releasing the active
component in a dosage form one may introduce a number of
difficulties. Quite often the problem that arises because
of these difficulties is not immediately apparent and may
cause a great deal of consternation. Very often the data
appears to be implausable.

Let us consider the set of circumstances where one
usually obtains an assay value on an active as a raw material
of 100% and usually obtains an assay value on this same
active component in its dosage form of either 98% or 102%.
When this occurs we normally conclude that:

1. We have a problem with the analytical assay method
   for the dosage form yielding a 2% lower value
   because of its inability to extract all the active
   or yielding a 2% higher value because of some sort
   of formulation interference.

or 2. The pharmaceutical process has broken down and the
   dosage form was not prepared properly.

There is a third possibility which, as we stated earlier, is
often overlooked, and that is that there is a different
criteria for the active component as a raw material vs. the
active component in the dosage form. If one is using different analytical methods for the raw material and the dosage
form and thus is measuring different portions of the molecule
it is quite possible to obtain 2% differences in assay values.

I recall from my own experience a very good example of
this problem. This problem revolved around the analysis of
ethinyl estradiol. At the time that we encountered this
particular problem the USP monograph for ethinyl estradiol
did not have an assay method and consequently we developed
an ethinyl titration procedure for testing and releasing the
raw material. In the dosage form, which was a combination
progestational-estrogenic tablet, the ethinyl estradiol was
assayed by an automated fluorometric method. Analysis of
the raw material would routinely yield an assay value of 100%
whereas the analysis of the tablets would yield assay values
of 103 to 104%. At first we blamed the pharmaceutical process. Then we looked for excipient interference. It was
only after study of these two possibilities that it was
realized that the raw material assay technique does not use
an analytical reference standard whereas the dosage form
method requires a reference standard. The reference standard
we were using was a USP reference standard material. However
upon investigation it was determined that this material
assayed 3% lower than a material which we purified. Subsequent work positively established that the USP reference
standard material had 3% impurity and this material was
destroyed by the USP office. If we had been able to assay
the ethinyl estradiol raw material and dosage form using the
ethinyl titration procedure we would not have encountered
this problem. Further if we had used the fluorometric assay
procedure for analysis of the raw material as well as the
dosage form the problem would have been much more apparent
to us. This example represents only one way in which the use
of different methods for raw material and dosage form may
cause difficulties. There are, of course, other ways most of
which are related to the portion of the molecule analyzed.

Although problems are introduced because different
analytical methods are used for the raw material and the
dosage form one must recognize that often different methods
are required. That is the raw material assay method may not
have the requisite sensitivity and selectivity to be used in

the dosage form. This being the case one would have to
obtain a more selective and specific assay method for the
dosage form. From this point of view then it is reasonable
to use different methods but one must maintain an awareness
that such a difference can create problems.

ANALYTICAL METHODS FOR STABILITY SAMPLES - WHAT SHOULD
THEY TELL YOU?

Bernard Z. Senkowski

Hoffmann-La Roche Inc.

Nutley, New Jersey

Now that the previous speaker has discussed my topic,
I think I'd like to just enjoy the program.

Now to get serious: I'd like to mention that I am
Bernard Senkowski from Hoffmann-La Roche and my topic,
according to the card, is "Analytical Methods For Stability
Samples - What Should They Tell You?". May I read very
quickly the ground rules that one should consider if one
is in the pharmaceutical business and is thinking about
the stability of products. This subject is covered under
something that Dan Banes, I am sure, is quite familiar with;
it's called Section 133.13 of the current Good Manufacturing Practices. It states that, "there shall be assurance
of the stability of finished drug products; the stability
shall be determined by reliable, meaningful and specific
methods determined on products in the same container
closure systems in which they are marketed. The stability
is determined on any drug product that is to be reconstituted at the time of dispensing as directed on its labeling
as well as on the reconstituted product recorded and maintained in such manner that the stability data may be
utilized in establishing expiration dates".

When one looks at the stability of a substance or of
a dosage form, the same final goal is sought, perhaps by
different techniques. I would not agree with the former
panelist that one needs to have greater specificity for the
dosage form than the drug substance. I think inherently

one uses the tools that are available to characterize the
composition and the purity of the substance or the substance in the dosage form. In addition, it is important
to note whether there are physical changes occurring in
either case. This is important since physical changes can
affect important parmeters of the compound or dosage form
such as dissolution or bioavailability. Considering stability testing and analytical methods, I don't think we yet
have the universal tool which would permit dropping the
tablet in one end of the machine and find that we have an
answer that defines the dosage form or the drug substance
completely. We can of course use the tools that are available and those that are being developed which may permit us
to better obtain information about the compounds and dosage
forms that we are dealing with. For example, one can
measure the infrared spectrum very nicely on a dosage form.
One can also measure the infrared spectrum on a drug
substance. But, unless you have a functional group that
is capable of producing an absorption band in sufficient
intensity that you can detect low levels of by-products
without interference, this tool in itself is not adequate.
Gas-liquid chromatography is another tool that may be used
in either case. If the retention time is defined correctly,
then indeed this technique is a specific method of analysis.
But, here again, the method in itself is not adequate to
define the compound, so you connect a mass spectrometer to
the gas chromatograph. Yet all these tools at our disposal
do not give us 100% assurance.

What do we do about this dilemma of never knowing the
absolute purity of a dosage form or drug substance? There
are several approaches that one may take to obtain added
assurance of purity. For example, I think it's important
that one use accelerated conditions, drastic conditions to
determine how a molecule comes apart. If you know how the
molecule is synthesized then this information gives you
some indication as to what you might find if you expose
the substance or dosage form to adverse conditions. Once
the molecule is subjected to drastic conditions, you then
look to see how many pieces you can find. The question is,
"Do you look at the main component that's left, or do you
look for some of the small pieces that are being formed?".
In our experience, it is extremely valuable to look for the
small pieces that are being formed. If, in the course of
your studies, you know the degradation mode or the instability mode of a molecule, then you can use a tool such as

polarography or perhaps ultraviolet spectrophotometry to
determine the content and the stability of the substance
in a dosage form. The data that one accumulates on the raw
material is invaluable in guiding the chemist as to the
techniques and tools that will be applied in studying the
dosage form.

Analytical methods - I think you are all aware of the
analytical techniques available to us, such as the various
types of spectroscopy, chromatography, polarography, etc.
It is important in the study of the compound to utilize
those analytical methods that will give you the most information about stability of the drug substance or the dosage
form and will permit you to make the logical decision as
to the length of time this material may be used either in
production or by the consumer.

TESTS TO MONITOR THE ACCEPTABILITY OF RAW MATERIALS--
ARE THEY ADEQUATE?

Anthony J. Taraszka

Director of Control, The Upjohn Company

Kalamazoo, Michigan

In order to answer this basic question one must determine (1) the tests in question and (2) the end use of the
raw material.

General test methods such as those which are found in
the compendia are usually, but not always, adequate to
describe the identity and purity of the raw material. Compendial tests are designed to provide standards for broad
pharmaceutical usage in a wide variety of products, produced
by a number of different processes. These tests were not
designed, nor are they adequate, for describing the raw
material characteristics required for certain end product
uses and/or process methods. A major inadequacy of compendial standards is the use of colorimetric identification
tests in the official monographs for a number of drug
substances. The need for modern infrared identification
tests and, where necessary, X-ray test data is rather
obvious and will not be discussed further.

The remainder of my remarks will be confined to the
additional tests which are required to produce high quality
finished products. These tests by the very nature of their
unique applicability are not, and probably should not, be
incorporated into compendial monographs. These standards
and test methods are used to define particular characteristics

313

which a raw material must possess for use in a particular
product produced by a well defined, unique process and
utilizing a specific formulation.

What type of tests are we talking about? First, I
would like to consider tests which are used to define the
acceptability of raw materials for a specific end use of
the product. For example, consider the tests for color in
solution, or the color of raw materials in general. Specifications such as this may be very important for esthetic
purposes if the final product is a solution; but it may not
be as important if the material is used in an ointment or
as a small percentage of the total ingredients in a tablet,
powder, or feed additive. Particle size distribution is
another characteristic which requires rather individualized
attention. Particle size characteristics may be quite
important in determining the final characteristics of tablets
and suspensions but would only be an aid in the processing
of solution formulations.

One also must consider if specific requirements are
necessary in order to define the acceptable content of
foreign particulate matter in raw materials. I say "acceptable limit of foreign particulate matter" since a limit of
"none" is impossible to comprehend. Processing equipment
wears out by friction; therefore, some "equipment" may be
present in the raw material. Given infinitely precise
methodology for determining particulate matter content
would certainly increase the probability of finding these
extraneous materials. While the objective is to reduce the
quantity that is present, we have to admit that there is
some minute quantity which is present due to the manufacturing process. The acceptable level of total particulate
matter and the specific types of particulate matter depend
on the final product usage. For example, if you consider
the content of extraneous silica in magnesium stearate to
be used in a tablet formulation there would be little concern.
However, the presence of metallic particles in raw materials,
particularly if these are to be used in solutions which
contain amines, may lead to either product discoloration or
deterioration.

Another set of distinctive standards are those which are used to define raw materials for use in specific processes. As an example I have chosen tablet formulations produced by the direct compaction route in contrast to the more traditional granulation process. One can immediately grasp that moisture content of the raw materials becomes more important in the dry direct compaction process than it is in the granulation process. A commonly used tablet excipient, such as calcium lactate, might need a lower moisture content than is allowable by the compendial specifications for use in direct compaction processing. Flowability characteristics are much more critical in the direct compaction process. The differentiation between spray-dried and powdered lactose is important in the processing, yet both materials meet compendial specifications. Last but not least, surface area or particle size distribution of lubricants, such as magnesium stearate, are important considerations in formulation design. Whereas one particle size or surface area may be acceptable for a particular process, an entirely different standard is required for a different process.

The reply to the basic question on raw material test standards is both yes and no: Yes, they are adequate to define materials for a broad range of pharmaceutical usage; No, since they may require additional tests for a particular process or use. In all probability these additional tests should not be included in compendial monographs.

Automated Analysis

Edited by **Sut Ahuja**

AUTOMATION IN MICROCHEMISTRY

AUTOMATIC C-H-N ANALYZER CASE HISTORY

Grant M. Gustin

Norwich Pharmacal Company

Norwich, New York

This is a five-year case history involving the combined use of the Perkin-Elmer CHN analyzer Model 240, the Infotronics digital voltmeter Model 30, and the Wang calculator Model 360 with programmer. While the details to be reported here are of course explicit to the instruments just mentioned, it is hoped also that many of the ideas presented are applicable to instrumentation in general. Some of the details have been reported in Perkin-Elmer's house organ "Instrument News" in 1968, Vol. 19, No. 1.

We selected the Perkin-Elmer analyzer to perform our CHN determinations for several reasons. First, the analyzer's combustion chemistry of pure oxygen surrounding the sample and sample weight range of 2-3 milligrams were similar to our existing techniques in which we have already established confidence. Second, the DC output signals from the detectors of the analyzer could be displayed as easily by a digital voltmeter as by a strip chart recorder. We believed the accuracy and precision required of this analytical system was greater than that available by a strip chart recorder and felt the additional cost of the digital voltmeter was justified for this purpose.

From a managerial point of view, our heavy workload being equally divided between the two classical methods of Pregl and Dumas also favored this single CHN analytical system requiring one less operator. What was just as important to consider, but difficult to ascertain, was the

additional expertise required to maintain satisfactorily
such a complicated analytical system. It is hoped that
this presentation will answer some of the questions relating to this problem.

Perhaps the most difficult decision came in selecting
the right digital voltmeter, what with the confusing variety of options available to the purchaser. We selected one
with a full scale range of 10-12 millivolts so that all
signals would be within full scale without attenuation.
It has a sensitivity of less than one microvolt, and a four-
second sampling period, or gate as electronics men say.

A later need arose for simpler data handling techniques
since a single analysis for all three elements involved the
manipulation of 8 raw data numbers and 6 constants. The
more economical electronic calculators, even those with a
little bit of programming and storage features, appeared
no more satisfactory than mechanical calculators already
in use. However, the Wang Model 360 and programmer had
programming and storage facilities sufficient to provide
the data processing simplicity we desired.

UPGRADING TECHNIQUES

The digital voltmeter (DVM) has two modes of printing:
single and repetitive sampling. Using the DVM on the repetitive mode of operation, commanded from the analyzer,
gives a progressive display of the CHN signals just as

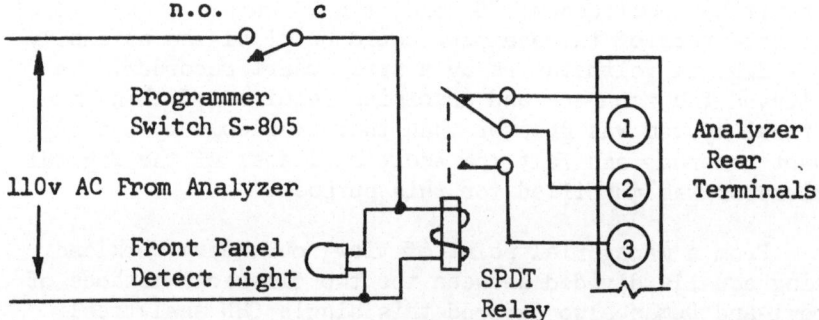

Fig. 1. Detect Switch for remote switching of the digital voltmeter "repetitive" print command using 110 v AC
power from Analyzer.

with the strip chart recorder. The analyzer's detect command switch numbered S-805 activates a 110 volt relay from
the analyzer's 110 volt system, the relay being in parallel
with the detect light as shown in Figure 1. The relay's
single pole double throw switch is wired in turn to the
digital voltmeter which supplies its own voltage over this
circuitry for remote repetitive printing.

Our reagents in the combustion tube, shown in Figure 2,
are all supported on chromosorb of the common grade-P, un-
washed, 30-60 mesh size. These combustion reagents, previously developed for our former Pregl combustion method,
proved entirely satisfactory in the analyzer.

A few helpful hints on the preparation of these reagents are in order. Silver oxide was precipitated from
solutions of silver nitrate and sodium hydroxide, the hydroxide being in slight excess. Cobalt oxide was first
prepared as cobalt oxalate precipitated from solutions of
cobalt chloride and oxalic acid. This oxalate powder was
then decomposed in a furnace starting around 250-300°C
with gradual increase to 450°C as the decomposition pro-
gressed. It was stirred frequently to facilitate the decomposition and avoid the formation of dense sintered
particles. Silver tungstate was precipitated from silver
nitrate and sodium tungstate. Silver vanadate was precipitated from solutions of silver nitrate and sodium
vanadate. Magnesium oxide was the commercial heavy grade
powder form. Thorough rinsing of the precipitates 4 to 6
times by decantation was essential with large aliquots of
distilled water. A rinse with acetone, then with chloroform followed before filtering in a Buchner funnel. The

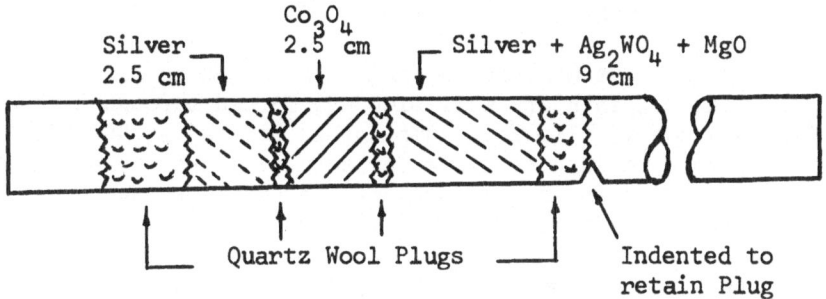

Fig. 2. Combustion Tube Packing.

Table 1

Ratios of Reagents Supported on Chromosorb
for Combustion Tube Packing
and Mixed as Catalyst for Addition to Samples

| Reagents | Combustion Tube Packings | | | Sample Catalyst |
	Silver	Co_3O_4	$Ag/Ag_2WO_4/MgO$	
Ag_2O	3		1	
Co_3O_4		1		2
Ag_2WO_4			1	1
MgO			2	
$Ag_2V_2O_7$				1
Chromosorb*	2	2	6	

*Moistened in the ratio of 3 to 1 of water (w/w).

organic solvent washes were important to keep the precipitates as fine powders. Otherwise, they sometimes became
densely packed hard granules. To get the finely powdered
reagents to coat the chromosorb in the proportions shown
in Table 1, chromosorb was first blended with distilled
water in a beaker in the proportions of 1 part water to
3 parts chromosorb by weight. The reagent was then added
and tumbled in the beaker with a fast swirling wrist motion
until uniformly coated. The pelleted reagents were dried,
then muffled in a furnace at 400-500°C.

The catalyst mixture in Table 1 is added to all samples. The reagents are well blended, forced through a fine
60 mesh sieve to eliminate lumps, then muffled at 400-450°C
in a clean porcelain crucible for several hours. The catalyst mix is stored in a tightly capped bottle to avoid atmospheric contamination.

Our types of samples severely attacked the platinum
sample boats so we use the disposable type aluminum boats
exclusively. These boats and the catalyst, in turn, both

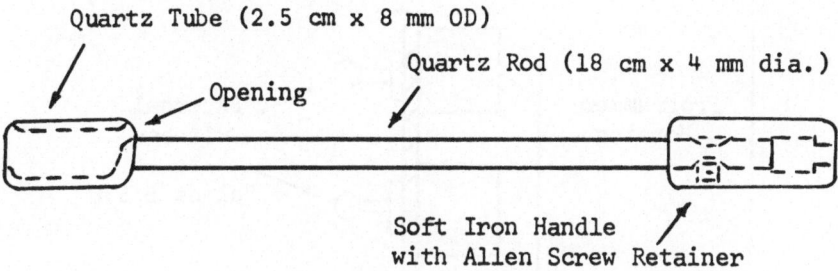

Fig. 3. Simplified Quartz Ladle with iron handle.

cause rapid deterioration of the sample ladle. The ladles have a short life expectancy of 90-120 analyses.

To minimize ladle costs, the simplified ladle in Figure 3 is easily fabricated in the laboratory. The original glass handle containing a small iron bar was replaced with an iron handle in which the long glass rod was secured by an allen screw. A hole was put in the other end of the bar sufficient for a hook to remove the ladle from the combustion tube. The sample tube was easily replaced on the long glass rod when deteriorated beyond further use.

Nighttime helium pressure was reduced from the normal operating pressure of 18 lb/in^2 down to 12 lb/in^2. This maintains the necessary protective flow of helium for the detectors and at the same time conserves the helium supply.

Several of our upgrading techniques were instituted mainly to simplify maintenance and operation, and to minimize the down time of the instrument.

Since the index wheel of the programmer is frequently used manually in advancing the analyzer's cycle to make occasional routine maintenance changes of the combustion and reduction tubes, it was remounted outside the instrument at right angles to the programmer as shown in Figure 4, with a right-angle gear drive connecting the wheel to the programmer shaft. Also two switches were mounted on the side of the analyzer near the wheel for the purpose of controlling valves B and E, with the electrical connections shown in Figure 5.

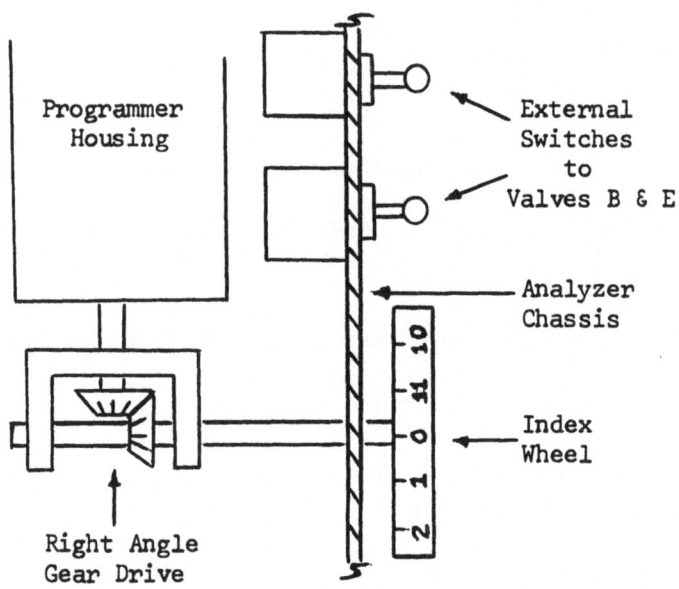

Fig. 4. Programmer Index Wheel remounted outside Analyzer chassis along with two switches for manually controlling oxygen flow (valve B) and helium purge (valve E) to
the atmosphere.

During diagnostic studies of the helium blank, it is
necessary to stop the flow of oxygen through valve B. With
great inconvenience, the valve had to be disconnected underneath the furnace tray next to exposed electrically hot connections. The external switch to valve B eliminates this
hazard and inconvenience.

When changing the combustion or reduction tube, the
external switch to valve E is used to reduce the instrument
down-time. The programmer wheel is first advanced manually
to the number one position whereupon helium flows back
through the mixing chamber to the atmosphere at valve E, as
well as forward through the combustion and reduction tubes
to valve E. Next the connector between the glass tubes is
removed. Immediately after this seal is broken, valve E is
turned off with the external switch that forces helium, previously backflushing through valve E, farther back through
the reduction tube to the atmospheric opening at the broken
seal. A new reduction tube is pushed rapidly through the

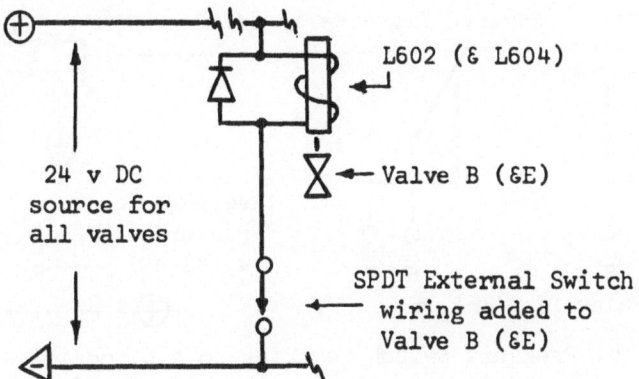

Fig. 5. Schematic wiring of manual switches to the oxygen valve (B) and purge valve (E) to the atmosphere.

hot furnace into the rigid end fitting, whereupon helium immediately flushes air from the tube before its copper packing becomes hot enough to become oxidized. If the combustion tube is changed at this time, oxygen is purged through it to expedite the conditioning of the packing. This is accomplished by simply advancing the timer wheel to position 2 whereupon oxygen flows from valve B, through the combustion tube to the atmosphere. Oxygen is allowed to condition the tube for 5 minutes, then is purged from the tube with helium by manually advancing the timer wheel again to the number 1 position. Now the connector between the tubes is secured and valve E is immediately switched back on which allows helium to purge through it in the normal manner. After 2 to 3 minutes, the timer wheel can be advanced to the zero or standby position. A single unweighed sample analysis is the only further conditioning of the combustion or reduction tube required.

The analyzer and the digital voltmeter each have a component that periodically fails and requires routine expensive replacements. A specially designed pressure switch in the analyzer, with a calibrated pressure inside, slowly loses this pressure. Also the switch inside sometimes becomes dirty and malfunctions, causing a slave relay and helium valve A to chatter. After replacing 3 of these pressure siwtches in less than 3 years, we rebuilt the faulty ones with difficulty and not with complete success; leakage continues. The malfunctioning switch problem was solved by rewiring it in the normally open position shown in Figure 6

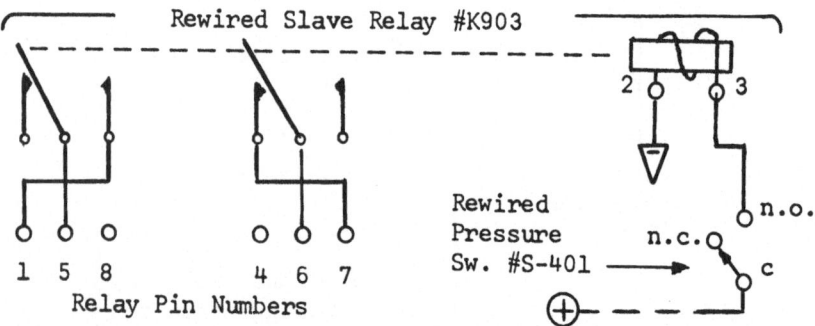

Fig. 6. Pressure Switch rewiring to N.O. position and wiring changes required on slave relay to maintain normal switching functions.

instead of the normally closed position. The slave relay K-903, which responds to this switch, now functions in the reverse mode of normally de-energized. This involved a simple task of removing the plastic cover of the relay, de-soldering wires at the normally open and normally closed positions, and reversing them. Now the pressure switch and slave relay operate only for a few minutes during each analysis, then remain off for the rest of the day.

The digital voltmeter's chronic replacement problem is an electromechanical chopper which fails faithfully every 8-10 months. A spare is kept on hand and plugged in when necessary. Pulling the old one out requires first the removal of screws securing it from underneath. Newer digital voltmeter systems have managed to eliminate this device; a worthwhile consideration when comparing more expensive systems. There is a reason for the additional expense.

INNOVATIONS

Our first innovation was the substitution of an electric solenoid coil in place of a permanent magnet for moving the ladle into the combustion furnace. Moving the ladle into the furnace must be done quickly and completely without losing control. While this manipulation is not difficult, a careless failure may result in a flashback of the sample into the cool part of the combustion tube, and a delay in further analyses until the ladle and combustion tube are

thoroughly burned clean and reconditioned. A 100 volt coil,
manually controlled by a push button switch, worked excep-
tionally well. A 24 volt coil was found inadequate. A word
of caution - to protect the coil from overheating and burn
out, voltage to the coil must be brief, 3-5 seconds. This
is ample time to insert or withdraw the ladle.

The most time saving innovation was next added - an
automatic drive to move the solenoid for inserting and extracting the ladle. Kunz in South Africa also has done
this and published an article entitled "The Perkin-Elmer
Elemental Analyzer Made Automatic" in the Microchemical
Journal Vol. 13, pp. 463-66, in 1968. His device still
uses the permanent magnet. The inject and detect lights
provide the necessary electricity to activate relays that
turn on the motor drive and magnet at the appropriate times
for inserting the ladle, for withdrawing the solenoid from
the hot furnace area, and later for again advancing the
solenoid for extracting the ladle during the detect period.
This innovation, coupled with the digital voltmeter's complete control of the detection periods, completely frees
the technician from attending the instrument for 13-14 minutes after each sample change.

Later the automation of changing ladles was accomplished. This eliminates the delays between analyses when
the operator fails to return at the proper moment to reload.
Several ladles can now be loaded at the operator's convenience. This phase of automation does not reduce the sample
loading time required of an operator, but its purpose has
been mainly to allow the operator to use his time more
efficiently with other tasks so that they do not have to be
interrupted the instant the analyzer completes its cycle.
Without automatic loading, the analyzer has to remain idle
at the conclusion of an analysis until the operator returns.
The use of several different ladles did not vary the CHN
blanks significantly, but all ladles have to be precondi-
tioned each morning. They are burnt off in an auxiliary
furnace at 800-900°C for 5-10 minutes while being purged
with oxygen. The changer operates much like the bolt of a
rifle to remove a used ladle from the combustion tube.
Perkin-Elmer very nicely provided a spare cam, switch number S-806, in the programmer which goes on during the last
few seconds of the cycle. This was used to activate the
changer. After the changer pushes the old ladle aside with
a new one and inserts the new ladle into the combustion

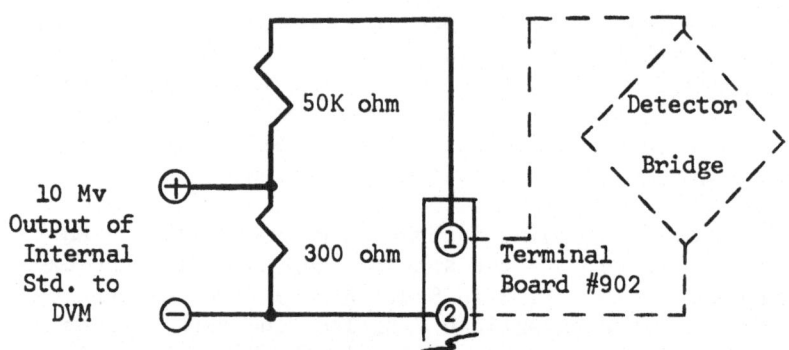

Fig. 7. Internal Voltage schematic wiring to same voltage as supplied to detector-bridges.

tube, sealing the tube at this time, the changer then com-
mands the analyzer to start.

The changer is idiot proof in one respect: it can be
started manually only when the analyzer cycle timer is in
the standby or off position. Once the analyzer has started,
the bolt cannot be withdrawn from the combustion tube, which
would let air into the system, either by accident or on purpose.

Our most remarkable innovation for maintaining consistent accuracy, yet the simplest to install, is an internal
voltage standard which is measured by the digital voltmeter
after each analysis. This printout reflects the performance
of the entire electronic system starting from the main voltage of the bridge networks, through the chopper and amplifier,
and finally to the printout of the digital voltmeter. Such
a standard would be just as useful for a strip chart output.

The internal voltage standard, shown in Figure 7, consists of two resistors in series connected to the same 16
volt power terminals to which the bridge detectors are attached. The resistors should be of the same high quality
as those used in the bridge networks. I used a 300 ohm resistor in series with 50,000 ohms so the voltage across the
small resistor, around 10 millivolts, would be approximately
80-90% of the DVM full scale. This voltage is then printed
5-6 times at the conclusion of each analysis and used in
calculating the results. The millivolt measurement of each
element, after the zero and blank subtractions, is divided

$$\text{Corrected Signal} = \frac{\text{Final Signal} - (\text{Zero} + \text{Blank})}{\text{Internal Voltage Measurement}}$$

Fig. 8. Internal Voltage Calibration correction as applied to one of the detector signals. Measured after each analysis, it is divided into the carbon, hydrogen, and nitrogen signals.

by this printout as shown in Figure 8 so the result becomes a relative measurement that is independent of electronic changes. The actual calibration voltage level is not important, but changes from this level are, which indicate the amount of correction required of the bridge signals.

This innovation hardly seemed justified at first, with months of little variation in the voltage standard. Before this correction, the analyst assumed that his recording system, after calibrating with standard samples, was maintaining the same performance through the rest of the day. He had no way of recognizing subtle drifting in amplifier performance as shown in Table 2. The same analyses with internal-voltage-correction improved the results with a 50% reduction in the relative standard deviation from 0.23% for standard calculations to 0.12%, as the digital voltmeter output drifted first below, then above full scale.

Even large sudden changes can be undetected. Only when the recorder signals were erratic did the analyst have sufficient cause to distrust his results. It took only one day of faulty chopper performance in the digital voltmeter to show the full value of this extra calculation effort as shown in Table 3. The deteriorating DVM gave useless output signals by standard calculations which meant valuable instrument/operator time wasted. But, with the same signals being divided by the internal-voltage-measurement recorded after each analysis, the results become entirely acceptable. Now it is possible to retain valuable analytical results at the same time faulty electronic performance is revealed. With this high degree of confidence in results, duplicate analyses are not routinely made which in effect doubles our work capacity.

Table 2

Effect on Carbon Results from Minor DVM Variations
During the Day without, and with
Internal Voltage Calibration Corrections

Analysis No.	DVM Deviation (%) from Full Scale	Carbon Deviation (Absolute) from Theory	
		Uncorrected	Internal Voltage Corrected
1	+.01	+.10	+.10
2	-.01	0	0
3	-.06	-.06	-.03
4	-.29	0	+.12
5	-.27	-.14	0
6	-.23	-.06	+.04
7	-.25	-.14	-.01
-			
-			
15	+.18	+.22	+.13
16	+.15	+.11	+.03

Statistical Analysis

Relative $\overline{X}$		99.96%	100.07%
Relative Standard Deviation		0.23%	0.12%

CALCULATION REVISIONS

Cleric et al., in Helv. Chim. Acta Vol. 46; pp. 2369-88, were able to show how carbon dioxide, water, and nitrogen in the helium system interferred with an accurate measurement of each of the individual elements and published correction charts for each element. We have not found the hydrogen and nitrogen measurements to be significantly affected by this phenomenon because of much greater dilution with helium at 2 atmospheres of pressure, but the carbon measurement is. The hydrogen error creates higher carbon results, while the nitrogen error creates lower carbon results. For standards and unknown compounds of similar relative content, these errors go unnoticed. But we had a rude awakening when we

Table 3

Recouping Useless Analytical Results from a Major DVM Failure
by Compensation with the Internal Voltage Calibration Signal

Compound	Signal Loss (%)	Standard Calculation			Internal Voltage Correction		
		C	H	N	C	H	N
Standard #1	.03	-.01	+.02	-.02	+.01	+.02	-.01
ANALYSES 2 HOURS LATER							
Standard #2	3.01	-1.82	-.19	-.49	+.06	-.05	+.11
Unknown #1	3.93	-2.57	-.08	-.70	-.27	+.06	-.13
Unknown #2	3.66	-2.21	-.20	-.62	+.27	+.07	-.10
Unknown #3	3.69	-2.58	-.24	-.70	-.09	+.03	-.18
Unknown #4	3.09	-2.16	-.10	-.17	-.29	0	+.03

(Absolute Error)

mg Carbon = mg Carbon - (A)(mg Hydrogen) + (B)(mg Nitrogen)
(corr.)   (uncorr.)

Constants: A = .0163; B = .0067

Fig. 9. Carbon Weight correction for water-nitrogen
effects that change carbon detector response.

tried some pure hydrocarbon samples containing 12-14% hydrogen which are 2 to 6 times more than our usual levels. We
then derived correction constants for multiplying the hydrogen and nitrogen weights, the results being applied to the
carbon weight as shown in Figure 9.

We found the hydrogen measurement to be greatly affected by the previous analysis because of an absorption phenomenon. It would be high or low depending on the water
remaining in the system. The system also seemed to retain
a basic amount of water, probably in the combustion tube
packing for the most part. We derived a correction formula,
shown in Figure 10, reflecting the effect of both the previous
analysis and the basic water retained in the system. Thus
when the previous analysis has been a blank, or the instrument has been idle for several hours, the hydrogen-of-the-
previous-analysis in the equation becomes zero. The hydrogen-
of-the-system and the constant D have been experimentally
established to give the tightest standard deviation. For
our system the constant D is equivalent to 0.011 and the
hydrogen-of-the-system equivalent to 0.25 mg.

$$\text{mg Hydrogen (corrected)} = H + \frac{D(H - Hp)}{H + Hp + Hs}$$

H = uncorrected mg Hydrogen

Hp = mg Hydrogen of previous analysis

Constants: Hs = 0.25; D = 0.011

Fig. 10. Hydrogen Weight correction for effect of
water retention from preceding analysis.

Table 4

Standard Format and Revised Format of Calculating Results
for 1971 AOAC Collaborative Study Compounds

Compound	Carbon Std.	Carbon Revised Calc.	Hydrogen Std.	Hydrogen Revised Calc.	Nitrogen Std.
Nicotinic Acid	+.12	+.14	+.03	+.01	-.03
	+.09	-.03	+.08	0	+.16
Stearic Acid	+.34	-.03	-.20	-.05	.00
	+.38	+.01	+.02	+.14	+.02
Acetanilide	+.05	-.01	+.06	-.03	-.08
	+.10	-.04	-.01	-.03	-.01
Methyl Palmitate	+.39	+.15	+.07	+.13	.00
	+.32	-.13	-.10	+.02	+.04
Ethyl Laurate	+.17	-.21	-.03	+.04	.00
	+.25	-.08	+.16	+.10	-.01

(Absolute Deviation)

Statistical Analysis

	Carbon Std.	Carbon Revised Calc.	Hydrogen Std.	Hydrogen Revised Calc.	Nitrogen Std.
Relative $\bar{X}$	100.30%	99.98%	100.14%	100.24%	100.01%
Relative Standard Deviation	0.16%	0.16%	1.20%	0.59%	0.57%

Our participation in the AOAC collaborative studies of
1971 clearly illustrated the standard calculation problems.
Table 4 shows the result of our collaboration using both the
standard format and our revised format of correcting for the
1) internal voltage standard, 2) the effects of nitrogen and
hydrogen on carbon, and 3) the effects of the previous analysis on the hydrogen. The revised format revealed two interesting and important statistical facts. Bias in the
accuracy of the carbon results improved dramatically from
100.30% by the standard format down to 99.98%; precision
remained the same at 0.16% (RSD). The opposite effect occurred with the hydrogen results; precision improved 50%
with a RSD change from 1.20% by the standard format down to
0.59%; bias in the accuracy increased slightly from 100.14
to 100.24%. The nitrogen calculation required no revisions.
The three non-nitrogenous materials reflected the excellent
reproducibility of the blank factor with results deviating
from -.01 to +.04% off theory of 0.00%.

What had been a simple method of calculating results
in the past now becomes an awesome chore with so many corrections. However, our Wang electronic calculator, with
punched IBM cards containing the programs for calculating
and correcting the measurements, provides us with a very
speedy output of the results. This programming feature has
allowed us to expand our service to include an empirical
balance with those results that do not agree with the chemist's theoretical values. This involves little extra effort
on our part while the results are still retained in the
electronic calculator. This empirical balance requires only
the change to program cards containing the empirical formula

$$\text{Carbon At. No.} = \text{Nitrogen At. No.} \times \frac{\% C}{\% N} \times \frac{\text{Nitrogen At. Wt.}}{\text{Carbon At. Wt.}}$$

$$\text{Hydrogen At. No.} = \text{Nitrogen At. No.} \times \frac{\% H}{\% N} \times \frac{\text{Nitrogen At. Wt.}}{\text{Hydrogen At. Wt.}}$$

$$\text{Molecular Weight} = \frac{\text{Carbon At. No.} \times \text{Carbon At. Wt.} \times 100}{\% \text{Carbon}}$$

Fig. 11. Carbon-Hydrogen-Nitrogen Empirical Balance
calculation and postulated molecular weight from analytical
results, assuming whole atomic numbers for nitrogen.

Table 5

Typical Deductions from Empirical Ratio Calculation of Results not Agreeing with Theory

	C	H	N	Molecular Weight	Deductions
Theory	19	16	5	398	Recovered starting material
Found	12.1	8.2	4	313	C-12, H-8, N-4, M.W. 311
Theory	7	8	4	212	Lost N-oxide form of oxygen
Found	7.0	7.9	4	197	
Theory	5	4	2	156	$+NO_2$
Found	5.1	2.9	3	205	
Theory	12	8	4	311	Chlorine lost – replaced by
Found	14.0	13.9	5	320	CH_3-N-CH_3 group
Theory	15	9	1	290	Changed triple bond to double bond
Found	15.3	10.3	1	334	and added a Cl and H atom

$$\text{Hydrogen At. No.} = \text{Carbon At. No.} \times \frac{\% \text{ H}}{\% \text{ C}} \times \frac{\text{Carbon At. Wt.}}{\text{Hydrogen At. Wt.}}$$

$$\text{Molecular Weight} = \frac{\text{Carbon At. No.} \times \text{Carbon At. Wt.} \times 100}{\% \text{ Carbon}}$$

Fig. 12. Carbon-Hydrogen Empirical Balance calculation and postulated molecular weight from analytical results having no nitrogen present, assuming whole atomic numbers for carbon.

calculations shown in Figures 11 and 12. The program for Figure 11 has been published in Wang Laboratory's house organ "The Programmer" Vol. 2, No. 5, pp. 8-10 (1968). It first makes an assumption the molecule contains one nitrogen atom. If the resulting carbon-hydrogen numbers are nearly whole numbers, an acceptable ratio between the three elements is established and the program continues to determine the molecular weight based on this assumption. If the first assumption is not valid, additional nitrogen numbers are tested by the program until an empirical balance is attained. If there is no nitrogen in the structure, an alternate program based on the calculations in Figure 12 is required. This program assumes the number of carbon atoms and displays the hydrogen number for the assumption made. It then calculates the molecular weight when a valid ratio of nearly whole numbers is attained.

The organic chemist's main interest is the empirical results. This service reveals to him immediately whether other structures are evident or just a simple further clean-up of the compound is all that is necessary. Examples of the revealing information from empirical ratio reporting are shown in Table 5.

In conclusion, I must say that we are extremely well pleased with the performance of our analyzer, digital voltmeter, and electronic calculator.

I hope this past five years of effort is just the beginning for us, not the end. I would like to see our equipment directly tied into computer facilities as some, I know, already have done.

METHODOLOGY PROBLEMS IN AUTOMATED ANALYSES

S. Ahuja

CIBA-GEIGY Corporation

Summit, New Jersey

Method development problems encountered in automated analysis generally stem from the limitations of the automated system. The problems encountered with each type of equipment are peculiar to that particular equipment. Only the methodology problems encountered with the AutoAnalyzer* will be discussed here. AutoAnalyzer is a modular system with versatility of handling a variety of operations such as homogenization or solubilization, dilution, dialysis or filtration, and liquid/liquid or liquid/solid extractions. Analytical determinations are made by colorimetry, spectrophotometry, etc., with the help of appropriate equipment. The following modules are generally used:

SOLIDprep
Liquid Sampler or Sampler II
Proportioning Pumps
Continuous Filter
Dialyzer
Mixing Coils
Automatic Digestor
Spectrophotometer, Colorimeter, etc.

The problems encountered and the solutions for each of these modules are described on the following pages.

-----

*Technicon Corporation, Tarrytown, N.Y.

SOLIDPREP

All comments are directed to the SOLIDprep I. Some of
these problems were noted by Technicon so they have developed a new SOLIDprep II, which is being currently evaluated
in our labs. However, SOLIDprep I has been and is still
largely used. Its operation is briefly described below.

1. The sample [solid, tablets, or capsule] is deposited in a cup placed on the turntable of the SOLIDprep. A
signal from a programmed sequence timer starts the turntable
and it rotates at a selected speed.

2. As the turntable rotates, each cup, at a certain
time interval, dumps the sample through a hopper into the
homogenizer. A stirrer [rotating at 10,000 rpm] dissolves
and/or suspends the sample in a preselected volume of the
solvent. The homogenization time can be varied as desired.

3. The homogenized sample is aspirated into the Auto-
Analyzer system and the stirrer operates at a reduced speed,
while the sample is being aspirated, to prevent settling
down of solid materials. The sampling time can be varied
as desired.

4. The unused sample is dumped into waste and the
homogenizer is washed with additional amount of solvent and
the waste is also dumped into waste. The wash time can also
be varied as desired.

SOLIDprep I was designed th handle up to 20 samples/
hour. This rate is controlled by appropriate selection of
gear for the driving mechanism. The sample rate can be
lowered to smaller number thus providing greater time interval for homogenization, sampling, and wash cycles. These
time intervals can also be altered by adjustment of the
related microswitches.

There are two major steps that determine the rate of
analyses:

1. Homogenization time - time necessary to solubilize
the sample.

2. Wash time - time necessary to wash the homogenizer
to eliminate sample to sample contamination.

More efficient homogenizers, such as that introduced
in the new SOLIDprep II, permits a faster rate of analyses.
Coupled with an efficient wash cycle, rate of analyses of
40/hour or more is advocated for this unit.

Corrosive solvents such as strong alkali or acid solutions, even alcohols at higher concentration could not be
used in SOLIDprep I because of deleterious effects on the
fittings of the SOLIDprep. By using teflon valves, one can
minimize these problems. We have been able to use solvents
such as 0.05N NaOH and 30% 3A alcohol.

In general the samples are presented as a whole unit.
In the case of tablets, the high speed of the stirrer during
homogenization can cause the tablet to eject from the hopper.
This can easily be prevented by the use of a plastic cover
or shield on top of the hopper. In the case of capsules,
however, different problems arise, such as a nonopening of
capsules or loss of powder during initial homogenization
stages. The problem of nonopening of capsules was resolved
by loosening the capsules slightly before placing them in
the cup. The loss of powder was prevented by arranging the
programming cycle as follows:

1. A portion of diluent is delivered.
2. The sample is dumped into the homogenizer.
3. The remainder of the diluent is delivered.
4. The stirrer starts only after all of the diluent
    and the sample are placed in the homogenizer.

Based on solubility of the compound being analyzed,
incomplete dissolution of the active ingredient can occur
when samples are analyzed at the rate of 20/hour. Furthermore, the homogenizer wash may be incomplete. A solution
to these problems of incomplete dissolution of sample and
homogenizer wash was found in the use of 4.5 - min. gear.

SAMPLER II

All of the problems mentioned so far can be circum-
vented by not using SOLIDprep; i.e., one can prehomogenize
or solubilize the sample and then use Sampler II for analytical work. This unit handles liquid samples so no homogenization is involved and thus a higher rate of analyses is
possible. Sampler II is similar to SOLIDprep in that you

place the liquid sample in a cup on the turntable. A mixer
keeps the material in suspension during sampling, and the
sample is aspirated from each cup, in turn, into the analytical manifold. If a wash cycle is necessary, each sample
cup is alternated with the wash fluid. Of course, this
slows down the rate of analysis.

It may be noted that samples can be presented to
Sampler II in the form of suspension. The suspended particles may be excipients or other insoluble material. If,
however, large particles are present in the suspension, it
may be desirable to prefilter this solution. Also, it is
important that the solutions be stirred during the sampling
step to insure uniformity of the sample.

MANIFOLD/PROPORTIONING PUMP

In this system, all reagents are pumped by peristaltic
action through various tubings. It is generally not advis-
able to overlap pumping tubes on top of each other--because
the pumping rate is affected and may not be very uniform.
In such cases, one is better off using two proportioning
pumps or Pump II.

With AutoAnalyzer, dilutions are carried out on the
basis of the size of internal diameter [i.d.] of the tubing.
In other words, the i.d. of the tubing determines the volume pumped/minute so the dilutions or other measurements
are carried out by the selection of tubes of the right i.d.
In general, the larger the i.d., the more accurate the
percent delivery. Tubings with an i.d. of 0.045 inches or
greater should be used in the manifold for good precision.

Acidflex tubings present different types of problems
such as "snaking". Some workers have found it difficult
to use acidflex tubings with Proportioning Pump II, because
it tends to wear them out faster or cause them to flatten
out; as a result, inaccuracies are introduced.

Another point to be remembered for the good base line
separation of peaks is that the manifold design should be
such that it clears itself quickly. This can be achieved
by using glass to glass connection [where possible] and
using minimal tubing lengths for various connections especially in the case of resampling line.

At times, problems arise due to hydrodynamics. For example, with one of our manifolds, the peaks were slightly skewed. The use of the time delay coil eliminated this problem [Figure 2]. To prove that this was actually the case, the peaks were recorded before and after the time delay coil was used in the analytical system.

CONTINUOUS FILTER/DIALYZER

Continuous Filter operates as follows: The sample flows continuously onto the moving filter paper--while slightly downstream clear liquid is aspirated into the manifold. Continuous Filter works well for most automated filtrations. However, it should be noted that because of its mode of operation, a large portion of the sample is rejected. This can be used to advantage if dilutions are necessary for analytical determinations. But when working with samples with small concentrations of active ingredient, this may prove to be a real problem.

While working with various release fluids [which are simulated gastric and intestinal fluids], containing phenmetrazine hydrochloride and various insolubilized tablet excipients, it was discovered that filter paper with very fine porosity could not be used as it led to aspirating of air into the resampling tube of the continuous filter. Schleicher and Schuell filter #410 proved satisfactory for this purpose [Figure 4].

In the case of samples which have very fine particles, the problem is reversed. Alternate means of filtration have to be found. Dialyzer with a suitable membrane can be used. Again, dilution does occur in the filtration process. Therefore, when working with lower concentrations of A.I., other ingenious steps are necessary.

MIXING COILS/AUTOMATIC DIGESTOR

Liquid/liquid extractions can be carried out with this automated system. For this purpose, beaded mixing coils are generally used. The layers are separated in a separator where the upper or lower layer can be processed further, as desired. Working with aqueous alkaline and chloroform systems, we found these coils unusable because they yielded very heavy emulsions which could not be resolved. This problem was resolved by using three mixing coils in a row along with two separators. The first separator provides the separation of chloroform from aqueous phase and in the second separator, chloroform is washed with water whereby emulsion problems are completely eliminated.

Another approach to extraction can also be used. This involves the use of Continuous Digestor. The use of Continuous Digestor can help by providing increased sensitivity and also eliminate emulsification problems. The Continuous Digestor is essentially a heated rotating helix into which organic and aqueous solvents are introduced at one end of the helix in such a proportion as to create thin layers of both of these phases. The organic phase is volatilized by heat and vacuum [which is applied on the other end of the helix] and the aqueous phase is aspirated into the automated system.

SPECTROPHOTOMETER/COLORIMETER

The solutions after processing are finally passed through the flow cell of a spectrophotometer or other suitable measuring device. For most analytical work, a colorimeter or spectrophotometer is used. Only the problems encountered with this technique of measurement will be discussed here. Flow cells generally used for this work are rectangular or tubular. Dimensions or cell thickness can be varied as desired. For example, one can use large cells such as 50 mm with the colorimeter or smaller cells such as 1 mm with a spectrophotometer. Naturally, the problems with each of these sizes are of a different nature. But let us restrict our discussion to 10-15 mm size, which are more commonly used. A tubular flow cell is desirable because it clears itself faster than rectangular cells; however, it is hard to align this cell in the light path and it is more apt to give base line fluctuations. Hence, the choice of geometry of the cell very much depends upon the objectives of the experiment. It is necessary to run standards--because the absorbance readings obtained vary from day to day--or as a matter of fact, during the same day. These changes may be due to changes in size of tubing, slight alteration of pumping rates, etc. Good precision is possible when standards are run along with the sample. It has been found desirable to run one standard before and one after each set of samples.

Once these problems have been resolved and a few other
details have been worked out, then one can develop reliable
automated analytical methods which afford considerable
savings in time and money. Furthermore, automated methods
help to generate a large amount of data which enables one
to study some interesting problems in the pharmaceutical
industry such as content uniformity and dissolution of drug
products. Some of the analytical methodologies thus developed are described below.

PHENYLBUTAZONE AND OXYPHENBUTAZONE

Phenylbutazone [3,5-dioxo-1,2-diphenyl-4-n-butylpyra-
zolidine] and oxyphenbutazone [1-phenyl-2[p-hydroxyphenyl]
-3,5-dioxo-4-n-butylpyrazolidine monohydrate] can be assayed by ultraviolet spectrophotometry [1-3]. The UV
methods provide the desired specificity and have been in
use for a long time in our laboratories. This procedure
was automated [Figure 1] with the help of Technicon Auto-
Analyzer [4] with the object of using it as a means of
analysis for the following four formulations:

1. Butazolidin [sugar coated] Tablets* [100 mg
   Phenylbutazone]

2. Butazolidin Alka Capsules* [100 mg Phenylbutazone]

3. Sterazolidin Capsules* [50 mg Phenylbutazone]

4. Tandearil Tablets* [100 mg Oxyphenbutazone]

The relative standard deviation with the automated method
is no greater than +2%. An average of the results obtained
for 10 capsules or tablets by the automated method compared
favorably with the results obtained on a composite sample
by the manual methods. The results for ten lots of each
of these products are summarized in Table I.

A very large number of individual tablets and capsules
have been analyzed by the automated methods, and the results

---

*Brand name, Pharmaceuticals Division, CIBA-GEIGY Corp.,
Ardsley, N.Y.

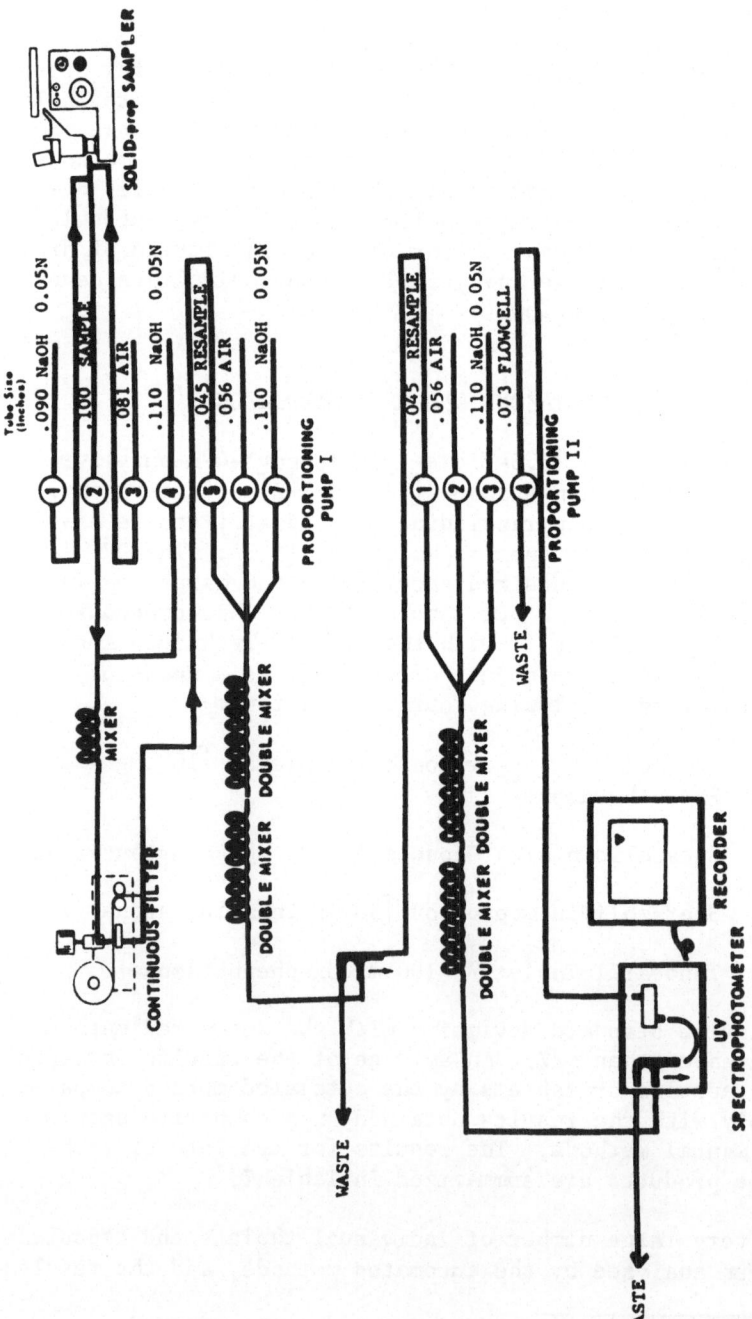

FIGURE 1 - Analytical Manifold for Phenylbutazone and Oxyphenbutazone - Homogenization
Solvent - 100 ml 0.05N NaOH; Analytical Determination - Phenylbutazone - 264 nm;
Oxyphenbutazone - 254 nm.

<u>TABLE I</u>

Summary of the Results for Phenylbutazone and Oxyphenbutazone Formulations

Formulations	Mg Phenylbutazone/Oxyphenbutazone		
	Automated	Manual	Deviation
Butazolidin Tablets [sugar coated]	97.8	98.0	±0.9
Butazolidin Alka Capsules	103.2	103.2	±0.6
Sterazolidin Capsules	52.2	51.9	±0.8
Tandearil Tablets	99.0	98.8	±0.8

obtained have very strongly substantiated the advantages of these methods for routine quality control and stability work.

IMIPRAMINE AND DESIPRAMINE

The automated methods of analyses [Figure 2] based on UV absorption of desipramine hydrochloride [10,11-dihydro-5-[3-methylaminopropyl]-5H-dibenz[b,f]azepine hydrochloride] and imipramine hydrochloride[10,11-dihydro-5-[3-dimethyl-aminopropyl]-5H-dibenz[b,f]azepine hydrochloride] were developed with the object of gaining information on inter-capsule and intertablet variations [5,6] on the following six dosage forms:

   a. Desipramine hydrochloride capsules
       10, 25, and 50 mg

   b. Imipramine hydrochloride tablets
       10, 25, and 50 mg

The precision of the automated methods was checked by run-ning standards and "synthetic" formulations. The relative standard deviation was found to be less than +2% for ten determinations. An average of the results obtained for ten capsules or tablets by the automated method compares favor-ably with the results obtained on the composite sample by the manual methods. The results on ten lots of these prod-ucts are summarized in Table II.

In order to study dosage variation, 590 individual capsules and 730 individual tablets were analyzed by these methods. The results are presented in Table III. The results very strongly suggest that in the case of these capsules, the content uniformity is at least as good, if not better than that of tablets, in the dosage range studied for these formulations. Needless to say, these tablets and capsules would have passed the content uni-formity test of the compendia [7,8].

PHENMETAZINE HYDROCHLORIDE

Automated analytical methods have been limited gener-ally to dissolution and/or extraction of active ingredient

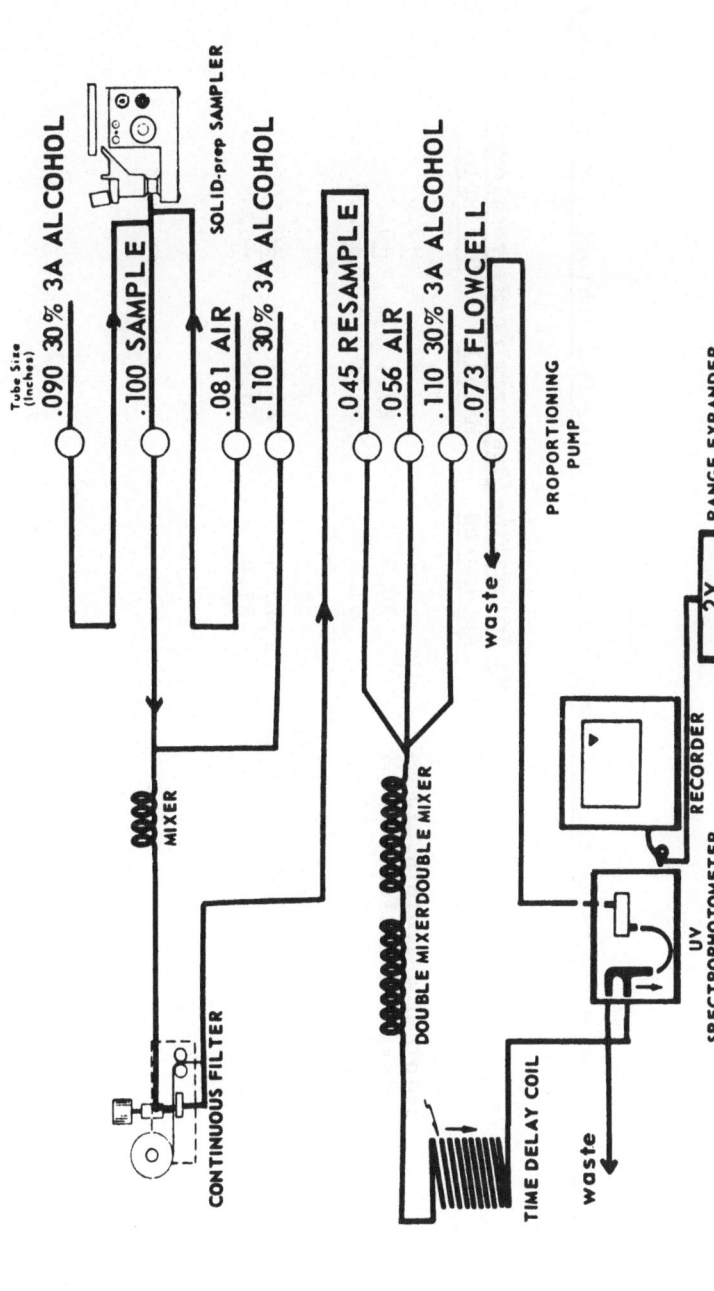

FIGURE 2 - Analytical Manifold for Imipramine and Desipramine - Homogenization Solvent - 125 ml of 30% 3A alcohol; Analytical Determination - Imipramine - 250 nm; Desipramine - 252 nm.

TABLE II

Summary of the Results for Imipramine and Desipramine Hydrochloride Formulations

Formulations	Mg Imipramine/Desipramine Hydrochloride		
	Automated	Manual	Deviation
Imipramine Tablets, 10 mg	9.79	9.82	±0.07
Imipramine Tablets, 25 mg	24.9	25.1	±0.33
Imipramine Tablets, 50 mg	51.0	51.7	±1.10
Desipramine Capsules, 10 mg	10.0	9.95	±0.14
Desipramine Capsules, 25 mg	25.4	24.9	±0.54
Desipramine Capsules, 50 mg	51.0	50.5	±0.63

TABLE III

Dosage Variation of Desipramine Capsules and Imipramine Tablets

Dosage mg.	% Capsules or Tablets Within Indicated Dosage					
	±5%		±10%		±15%	
	Tabs.	Caps.	Tabs.	Caps.	Tabs.	Caps.
10	57.4	65.0	90.3	96.7	99.4	100.0
25	79.6	67.8	99.2	94.1	100.0	100.0
50	82.9	88.7	99.0	100.0	100.0	100.0

from the formulation followed by filtration and dilution to
suitable concentration for analytical determinations.
Manual methods involving evaporation of the organic solvent,
followed by dissolution of the residue in aqueous solvent
have been considered difficult to automate. We attempted
to automate one such method [Figure 3] for Phenmetrazine
tablets which involves dual extraction [9].

As mentioned before, beaded mixing coils are generally
used for liquid/liquid extraction in automated methods.
For this work, these coils were found unusable because they
yielded very heavy emulsions which could not be resolved.
This problem was solved by using three mixing coils in a
row along with two separators. The first separator provides
the separation of chloroform from the aqueous phase. In the
second separator, chloroform is washed with water, whereby
emulsion problems are completely eliminated.

The second extraction step, involving re-extraction of
phenmetrazine into dilute hydrochloric acid [4 in 100] from
chloroform, was automated by the use of the continuous
digestor. This is accomplished in a heated rotating helix.
The chloroform and the dilute hydrochloric acid are introduced in such a proportion as to create thin layers of both
of these phases. The organic phase is volatilized by heat
and vacuum [which is applied on the other end], and the
aqueous acid phase is aspirated into the automated system
and processed as shown in the analytical manifold. The use
of the continuous digestor provides increased sensitivity
[10] and eliminates the emulsification problems commonly
encountered in manual liquid/liquid extraction methods.
The precision of the automated method was found to be good
[less than $\pm 2\%$ relative standard deviation] and the results
for ten different lots of tablets showed a deviation of only
± 0.3 mg for 25 mg tablets.

In order to study the in vitro release pattern of
phenmetrazine hydrochloride endurets [prolonged action
tablets] for quality control or stability studies, analyses
have to be performed on several tablets at the following
release intervals:

G_1 = first hour in simulated gastric fluid

I_2 = first two hour period in simulated intestinal
       fluid

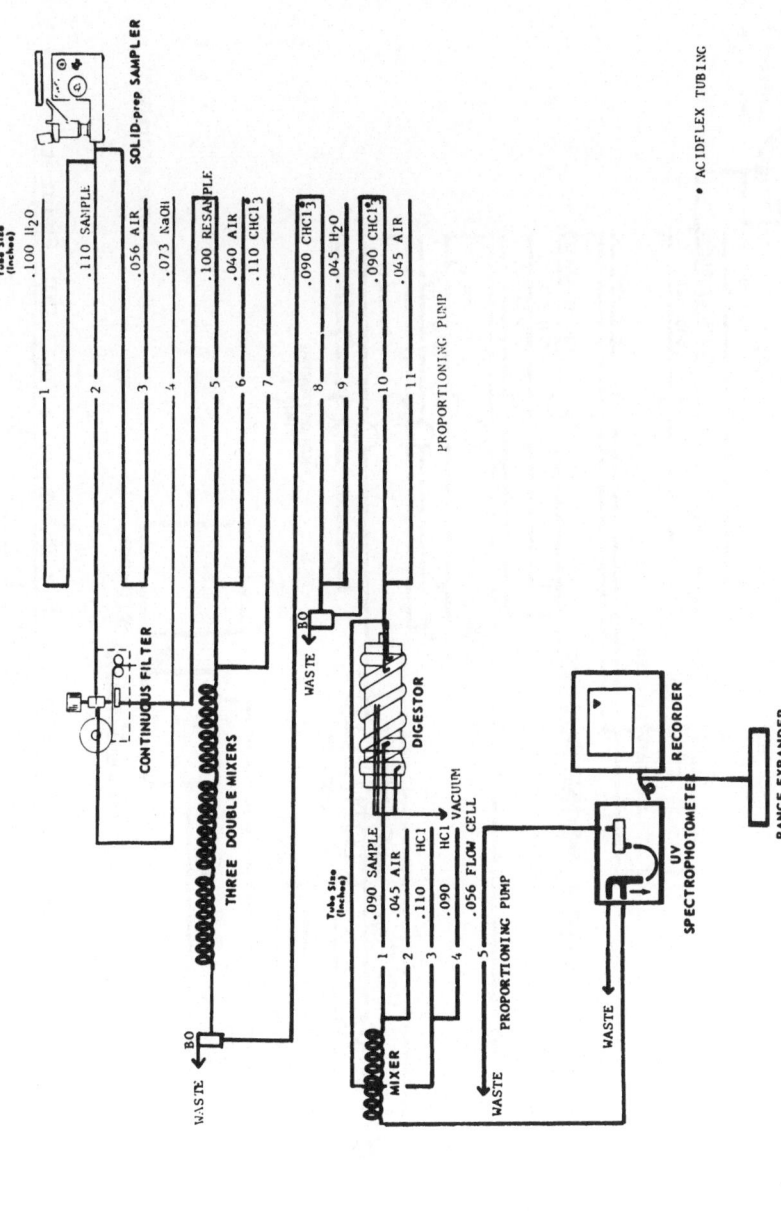

FIGURE 3 - Analytical Manifold for Phenmetrazine HCl Tablets - Homogenization Solvent - 100 ml water; Analytical Determination - 256 nm.

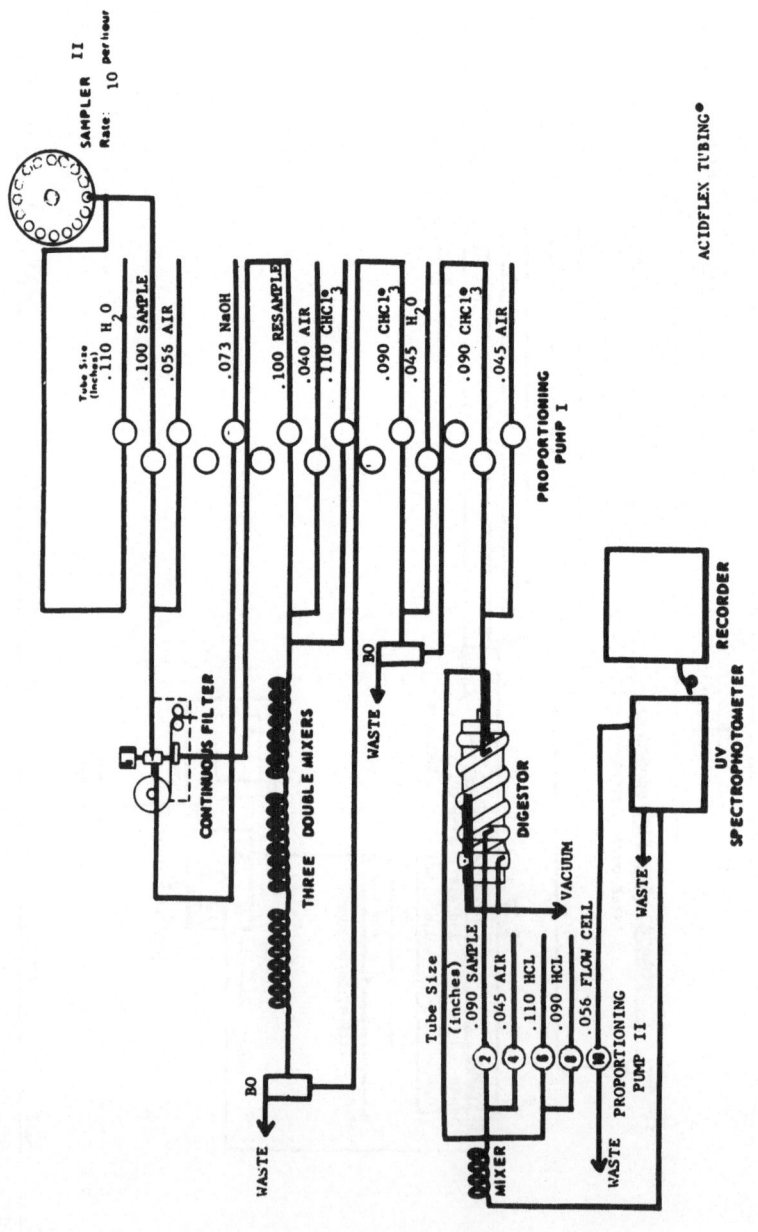

FIGURE 4 - Analytical Manifold for Phenmetrazine · HCl Endurets
Analytical Determination - 256 nm

TABLE IV

Summary of the Release Rate Results*

| Method | Mg Phenmetrazine Hydrochloride | | | | |
	Gastric	I_2	I_4	I_6	Residue
Automated	21.4	15.7	12.4	9.1	13.9
Manual	22.2	15.7	12.4	8.8	13.7

*Average of 10 samples

I_4 = second two hour period in simulated intestinal
fluid

I_6 = third two hour period in simulated intestinal
fluid

R = residue [nondisintegrated residual tablet]

For several years, a UV procedure involving extraction
methods has been used in our laboratories. The complexity
of automating this procedure is that one is dealing with
several different types of liquid samples of different composition, varying pH's and containing a wide range of concentrations of the active ingredient. The problem is
further complicated by the fact that the manual method of
analysis involves several extraction procedures. This
procedure was automated [Figure 4] so that all of the release fluids could be analyzed with the same manifold [11].

The precision of the automated method was checked by
running standard in simulated gastric and intestinal fluids
as well as dilute hydrochloric acid. The relative standard
deviation was found to be less than $\pm 2\%$. In general, the
results obtained by the automated method compare favorably
with those obtained manually [Table IV]. However, where
discrepancies were observed, it could be demonstrated that
they were due to emulsion problems encountered with the
manual method.

SUMMARY AND CONCLUSIONS

A thorough familiarization of the limitations of the
automated systems permits development of precise and reliable automated analytical methods which can afford considerable savings in time and money. Automated methods
permit detailed study of some interesting problems in the
pharmaceutical industry such as dissolution and content
uniformity. As the analytical parameters are set up rigidly, the product has to meet the same quality, otherwise
changes in the product show up as changes in the assay
values.

REFERENCES

[1] W. Fuchs, Muench. Med. Wochschr. 107, 1267 [1965].

[2] G. Guadagnini and R. Perego, Boll. Chim. Farm. 104, 815 [1965].

[3] M. Sahli and H. Zeigler, Arch. Pharm. Chem., 68, 186 [1961].

[4] S. Ahuja, C. Spitzer and F. R. Brofazi, Automation in Analytical Chemistry, Mediad Incorp., White Plains, N.Y., p. 467 [1968].

[5] S. Ahuja, C. Spitzer and F. R. Brofazi, Automation in Analytical Chemistry, Mediad Incorp., White Plains, N.Y., p. 439 [1968].

[6] S. Ahuja, C. Spitzer and F. R. Brofazi, J. Pharm. Sci. 57, 1979 [1968].

[7] The United States Pharmacopeia, XVIII, Mack Printing Co., Easton, Pa., p. 930 [1970].

[8] The National Formulary, XIII, Mack Printing Co., Easton, Pa., p. 798 [1970].

[9] S. Ahuja, C. Spitzer and F. R. Brofazi, J. Pharm. Sci., 59, 1833 [1970].

[10] B. Feller, W. A. Boyd, B. E. DiDario and A. Ferrari, Automation in Analytical Chemistry, Mediad Incorp., White Plains, N.Y., p. 206 [1967].

[11] S. Ahuja, C. Spitzer, Advances in Automated Analysis, Thurman Assoc., Miami, Florida, p. 227 [1971].

GROWTH PATH FOR COMPUTERS IN AUTOMATED ANALYSIS

Henderson Cole
IBM Thomas J. Watson Research Center
Yorktown Heights, New York, 10598

We are always interested in finding ways to increase our effectiveness as scientists. One way to do this is to see if there are better tools to help us in the jobs we face, or more effective ways of using the tools that we have. Attaching a computer to an instrument makes a powerful new instrument; any instrument, and almost any computer. A computer in itself is not an instrument, but when attached to any instrument it is. Thus, this is a possible technique for increasing our effectiveness which may apply to many disciplines. Rather than describe a specific instance, although I will give examples, I will try instead to give an overview of this type of growing activity, and try to state where I think it is heading. In this sense, this is somewhat of a selected version of what is going on in "Lab Automation".

For myself, this activity started some ten years ago when it was decided that certain exotic new materials might play a role in computer components, and the laboratory needed to build a capability in crystallography. Dr. Y. Okaya, a noted crystallographer, joined the group, and in

casting about for more people (hard to get), more
space, and more instruments, to achieve our goal,
the possibility of putting together a completely
computer controlled four axis x-ray diffractometer
looked as if it might be feasible, and was a
possible alternative way of doing the job. Dr.
Okaya, as a crystallographer, was a skilled
programmer; we also had a skilled instrumentation
man, F. Chambers, and myself. To make a long
story short, using an IBM 1620 computer in a
dedicated mode, with an in-house built interface,
and using parts of three different types of
spectrometers, we were able to "automate" the
taking of x-ray intensities from single crystals
for structure determination purposes. The result
was better than we had anticipated. By
conventional means, a group such as ours
ordinarily produced, say, two determinations a
year; with the automated diffractometer we
produced data at the rate of 12 analyses per year.
That is one a month, or a paper in every issue of
the journal. This was accomplished with less
investment than it would have taken to reach this
level of productivity by conventional means.
Also, the data analysis, as those of you who have
done such calculations know, were too big for the
little computer, so it was done on the larger
computer in the computing center. So we saw early
that the computerized instrument in itself was
only a part of the job facing the scientist. Our
job is to extract information about nature from
our experiments, not just the data. Obviously,
also, providing the figures for a publication a
month is quite a strain on the drafting group, so
we also turned to using the computer to produce
our curves, plots, and figures. (See, for
example, Figure 2).

As a result of this experience, some of us
turned our attention to encouraging others in the
lab to do likewise. Although not by the efforts
of a single group alone, there are now a wide
variety of instruments and experiments in the IBM
Research Laboratories making use of on-line
computers. The list includes:

1) Four axis x-ray diffractometer
2) Multiple x-ray spectrometers, for pole figures, phonon scattering, thin films, micro-probe, fluoresence, etc.
3) Particle backscatter
4) Mossbauer spectra
5) Multiple GLC's
6) UV spectrometer
7) Electron paramagnetic resonance
8) NMR, pulse and Fourier
9) Laser, other light scattering spectrometers
10) Quadrapole mass spectrometer
11) Activation analysis
12) Differential scanning calorimeter
13) IR spectrometer, including Fourier
14) Low resolution mass spectrometer
15) RGA
16) Crystal pullers, one with a TV-IR sensor
17) Microdensitometers
18) Curve follower
19) Tensile tester
20) High resolution stark spectrometer
21) Flash chemical kinetics
22) Magnetometer
23) Semiconductor device testing
24) Scanning electron diffraction
25) Scanning electron microscope
26) Lens design
27) Thin film thickness measurements
28) Non-dispersive x-ray detecting system

Many other examples of lab automation also exist in the literature now. Professor S. P. Perone, in a review article in Analytical Chemistry, 43, August 1971, lists 194 publications describing such efforts. Can we draw any generalizations yet? We will try to make some from here on out, but one thing which bothers us is that it is not yet any easier for the novice to get started. It is easier in the sense of learning what others have done, but not necessarily any easier in the doing it for yourself.

Enough has been done in this field to make it
clear that it is not a gadget-fad. What is its
value then? Primarily, its value lies in getting
the answers back sooner. If it is of value to you
to make a run, or do an experiment, is it of value
to have the results back sooner? Not just the
data, sometimes the data taking may actually be
slower, especially on a shared computer system,
but the answer is back sooner. Sometimes the
answer is back sooner because of experimental
techniques you wouldn't try without a computer,
such as Fourier IR or NMR; sometimes because the
better handling of the data in the computer
analysis program cancels the need to run a
corroborative experiment to cross check; but most
often it is just that the computer can analyze so
fast that calculated results are available almost
as soon as the data taking is finished. Often
results are available while the sample is still in
the instrument, or the instrument is still set up
for that type of run, so that immediate reruns are
possible if things look unusual. Thus, the
scientist is brought back much closer to the
experiment. Lab automation does not remove the
scientist from the experiment; on the contrary, it
requires his decision making capability much more
often than in the past.

Another encouraging sign that this is a step
in the right direction, is that very few ever
disconnect from the computer and go back to the
manual, or mechanized operation. This does not
mean that there are no complaints. After all, a
computer is just one more thing that can go wrong
and stop your experiment. But, when the computer
goes down, the effort is to fix the computer, not
remove it.

One important thing to keep in mind is that
any step towards automation should be taken in
terms of the larger job. Even if it is a small
step, it should be compatible with what you want
to do next. Figure 1 gives a somewhat gross
classification of the areas in which a computer
could be used.

SAMPLE HANDLING
INSTRUMENT SETUP

DATA ACQUISITION

    INSTRUMENT ATTACHMENT
    CHECK SETTINGS
    CALIBRATIONS
    DATA TAKING
    ERROR DETECTION
    FIRST PASS CALCULATIONS

INSTRUMENT CONTROL

DATA PROCESSING
DISPLAY RESULTS
LARGE CALCULATIONS
STORAGE OF DATA

INFORMATION RETREIVAL
FILE SEARCHES
REPORT PREPARATION

Figure 1.   Areas of Computer Use

Sample handling, and instrument setup, are usually the last items considered. The next few items are those most often identified with Lab automation, but it is more properly "instrument automation" or data taking. This involves interfacing, programming the instrument operation, calibrating, acquiring and cross-checking the data for validity, etc.

Often the close attention paid the instrument by the computer (i.e. calibrations) results in more significant data, which can by itself often justify the computer use. It is surprising how much you learn about previously unsuspected drifts in an instrument which can lead to reporting wrong, and often costly, results. Also, the results are more reproducible. The computer program may do the wrong things, only you can tell that, but it at least does the same thing each

time. Many subjective little choices made by
human operators are eliminated. Humans, of course
make these choices skillfully, and a computer
doesn't, which is why the initial programming to
get correct performance is often difficult.

      The data handling doesn't stop with the
data-taking. In fact, that is only the start.
The name of the game, really, is data analysis.
Information extraction is what we are really
after. That may involve storage of data for
historical comparisons later; file look-up;
merging results from many tests into a single
report for management. The data taking merely
feeds the real work. The strip chart recorder is
not the end of the world. In fact, we should rise
up and free ourselves from this limitation on our
view of our own work. It's the next steps which
are really important.

      What sort of implementations are possible?
Aside from off-line data logging into cards or
tape, with later computer processing, the
possibilities are symbolically
represented in Figure 2. The first is one-
instrument-one-computer-one program, usually from
one vendor. This is the best solution for a
well-defined, self contained operation. In fact,
this is very appealing to new users, because it
apparently by-passes the programming effort and
time required of them or their organizations.
But, it should be considered in terms of being
compatible with what will come next. Usually the
inevitable consequence of having a "computer" is
that one starts realizing what else "could be
done" if we "just modify the program". With more
than one program, you start a program library, and
then you want more storage (memory or bulk, disk
or tapes) to keep them on, and readily available
as needed.

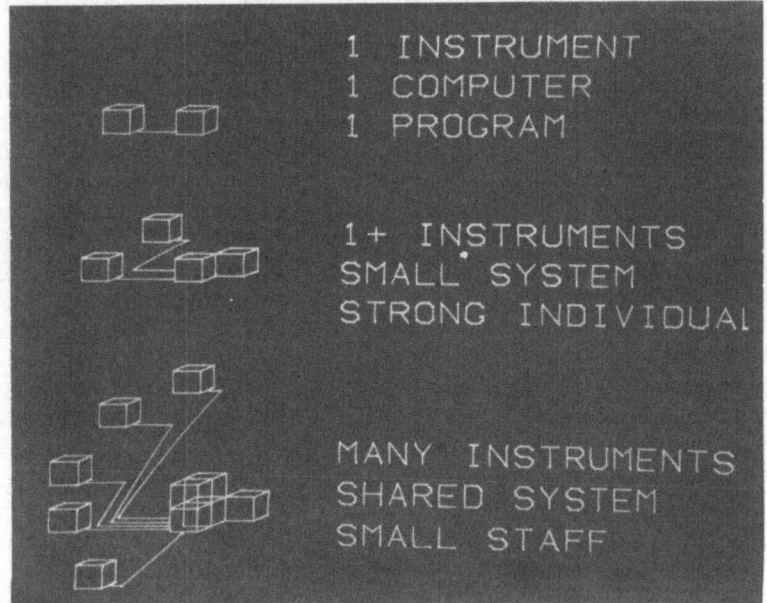

Figure 2.   Methods of Implementation

Often then the computer is too small, and being used for the "wrong" things in terms of its own capabilities. That is, compiling new programs is clearly better done in larger systems, not in small ones.

The next level is to start with a more complete computer "system", but perhaps still owned and operated, with a couple of instruments attached, by one "strong-willed" individual. This is a very popular way to operate today. More of the total range of jobs (Figure 1) can be done, including interactive graphical displays. But even here, one soon hears: "my computer is too small," but at the same time: "look how I am sharing it." A third approach is the larger shared system, usually installed as a "department machine." This usually requires a small staff to

run it (a hardware and a software person), and
gets into administrative problems. Its advantages
are that each user does have a larger computer
system when he needs it - most of the time, and
the system cost is shared over 10-20 users so that
the cost per user is quite acceptable. Its
disadvantages are those of keeping independent
users out of each other's hair, and preventing one
user from bringing the whole system down too
often. This approach usually gets installed
because one application needs it, and then the
other applications are brought on, since the
computer can also handle them. But, of course,
there are some users whose needs are not
compatible with those of the majority. If 10 GC's
are being handled, with 2, 3, data points per
second from each, and a user wants to use an
instrument and record 70 peaks in 5 seconds, that
is no problem, except that there may be a few
seconds gap in the data from the other
instruments. With data smoothing techniques, such
"lost" points are filled in (just as if the pen in
the recorder had not written for a short period,
and you draw the line in with a pencil). But, if
there are really significant interferences, who,
administratively, do you favor? The fact
is, of course, that computers are really very
fast compared to today's instruments, so several
operations can be interleaved; and as the
instruments become fast, they won't run for very
long periods. There are exceptions, of course,
such as the scanning electron microscopes, some
microprobes, or high resolution mass
spectrometers.

But, what is missing from Figure 2 is the
computing center. Where does it fit into the
picture? Too long, I believe, the scientists have
not regarded the computing center as a resource
that they can use. Most of our organizations have
a larger computer. We should take a look at it,
with our needs in mind. Some centers are now
offering "conversational" capability, where the
user can use the larger computer from a terminal
in his own area. This has been particularly
successful at the IBM Research Center, where

conversational capabilities, using APL, TSS, or
BASIC are available, and have made computers more
accessable and acceptable to many in the
laboratory. With this kind of support, we should
no longer feel constrained to the resources of the
little system in our own laboratories, but take
advantage of the larger resource to get more of
our total job done.

As an example of some of the detailed "system
analysis" questions that can arise, I would like
to discuss a computer controlled plate-reader, or
micro densitometer, to pick an "instrument" that
we are all "more familiar" with than perhaps any
other. Details of this application, both hardware
and software, were published in "Advances in X-ray
Analysis," Vol. 14, 1971.

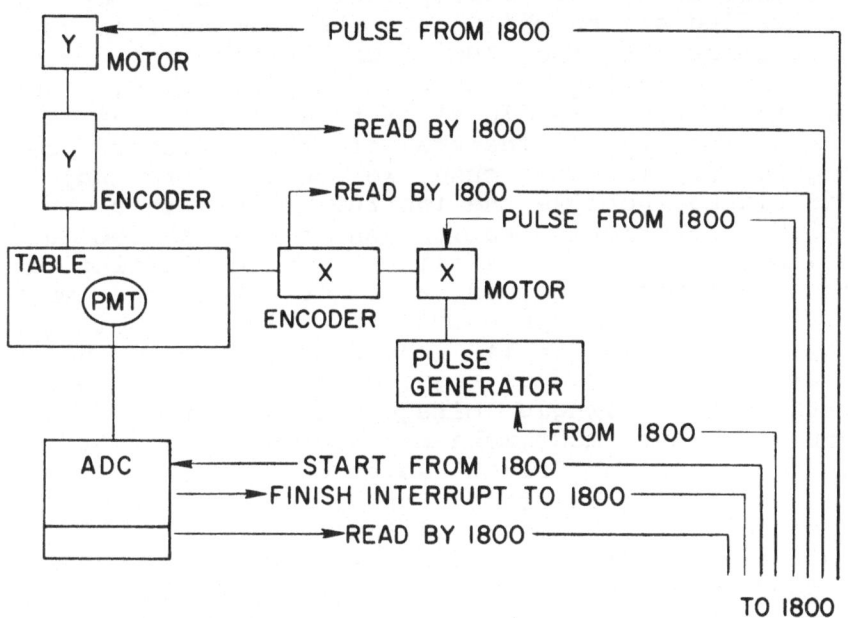

Figure 3. Microdensitometer Interfacing

In Figure 3 there is a block diagram of the elements of the instrument. It has both X and Y motion capability, but for the moment consider just the X-motion. A mass spec plate, or some other film recorded data, is placed on the instrument. it is set to its starting position, and as the table moves in the x-direction, the intensity of a transmitted light beam is detected by a photomultiplier tube, which puts out a voltage. Thus the instrument produces a voltage as a function of plate position (or time), which is usually put on a strip chart recorder; and thus it "simulates" the behavior of a majority of spectrometers. On an interfacing basis, we clearly need to digitize the signal voltage, the output of the PMT. This requires a digital voltmeter, or an Analog-to-Digital Converter (ADC). These can be bought, or they can be part of the computer. If a part of the computer, they are called the Analog Input feature. Most computers which have an Analog Input feature, have one ADC inside, but share its use by digitizing voltages from several instruments by use of a multiplexor. The multiplexor is a computer controlled switch which puts the voltage from your instrument into the ADC at the appropriate times. In this drawing we show a separate ADC. The digital output (readings) of this device goes into the Digital Input feature of the computer (see Figure 4). In this case, we chose this way of digitizing, because the instrument was in a very noisy electrical location, and there was too much voltage pick-up on the lines. By digitizing at the instrument, in this case, we eliminated much of the "noise" problems, because transmitting digital information is less susceptible to noise pickup. But it does require a multiwire cable, rather than a single voltage (signal) line. Thus, you see, one must make engineering "trade-offs"; there is no one "right" solution, although there is a "good" solution for each situation. A computer with a number of Analog Input points (a multiplexor, amplifiers, and an ADC), is however, a good way to get signals from a number of instruments digitized, or interfaced.

However, the really significant problem, is what procedure will be followed in the data-acquisition phase? One way is to start the plate moving, and let the computer catch the data on the fly, by using a clock to trigger off a reading of the ADC at equal time intervals. This procedure has to be used with time dependent experiments (such as a GC), but it is the most troublesome way to do things. What are "equal time intervals"? Every electronic device requires some time to function. Or again, what if the computer gets busy and misses some points? We could say then we don't care when the points were taken, if on the average about every second, but record the time when each reading was taken. We then add a digital clock to our "interface", and each time our trigger clock (our "interrupt", either external or provided internally by a program in the computer), is responded to, we record intensity and time, and adjust in the computer by interpolation to equal time intervals. But in this case, why bother? We don't really care about equal time intervals, but about equal steps along the length of the plate. In this case we choose then to use a "step plate, then measure" procedure. This puts the whole operation under control of the computer because the plate isn't advanced until the computer reads the last point. Always look for those occasions where an experiment can "stop and hold", and plan your strategy around those points. This is accomplished very easily here by a pulse stepping motor. Although we also attach an absolute reading Shaft Angle Encoder to each shaft, so at any time the computer can read the encoder to determine where the shaft (or table) really is. If we use a pulse stepping motor, the smallest step along the plate is determined by the lead screw pitch. What if the lead screw is not linear? We don't get equal steps along the plate then. If the steps are small enough this doesn't matter, if the motion is reproducible, because once the motion is calibrationd with a standard plate, the calibrate curve can be kept in the computer. In this case, the computer is able to enhance the functional capability of the instrument, without the necessity of mechanically building a better instrument. It often turns out that with the processing capability of the computer the data can be analyzed statistically in such a way as to increase the effectiveness of the instrument without redesigning the instrument itself. This brief, but perhaps too lengthy, discussion of some "system aspects" is meant to give a feeling of the challenge of the job of lab automation.

In Figure 4 we turn more to the computer part of the system.

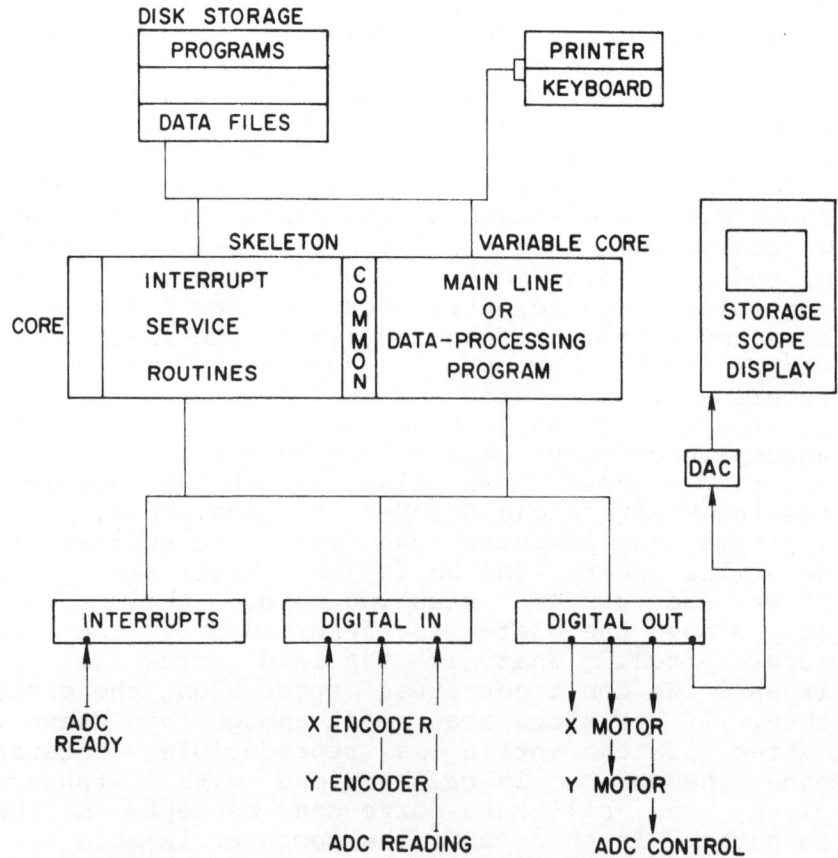

Figure 4. Shared Computer Features

Selected   features   of   interest   to   this
discussion are shown.   The computer   (an IBM 1800
class shared   computer)   has "front   end" features
which permits   "interfacing" the instrument,   such
as Interrupts, Digital Input, and Digital   Output.
Analog   Input and Analog   Output are shown in this
case as   separate   items   (ADC   and   DAC),   since,
functionally,   that is   the   way they need   to   be
thought of.   Internally,   in   the computer,   all
information is   represented digitally.   Now   some
small part of the core (memory)   must be dedicated
to   each   user, since   we want him   to   have   the
freedom to push   his own "start" button in his lab
at   any time,   without   previously having   had   to
schedule   when he   will   run.   This part   of   the
computer   must   always   be   "listening"   for   each
user's call.   The   computer   has   bulk   storage
(disks)   so   that each user's programs stay   in the
computer "system" all the time (but not in   core).
When the user signs   on, his   programs are brought
into   core, for   execution,   at the times they are
needed.   Thus,   when   several   users   are   in
operation,   their   programs,   or   parts   of   their
programs, and   their   data, are   being moved   from
core areas to   disk areas   (files).   Since certain
of their programs must   leave   information in core
for other   parts of   their   programs   to   work on,
there is   a   COMMON   area   in core   which   is   not
shipped in and out,   and each user keeps his   part
of this   area all the time.   Obviously, there must
be some   "master"   program in   the computer, which
keeps track of all   this activity, and that is the
"system" or monitor   program.   This   also stays in
core all   the time, and is usually provided by the
computer manufacturer.

From   the   user's   point   of   view,   what
structurally,   should   his programming   look like?
We try   to give a   feeling for this   in Figure   5.
These are   the   major "software steps"   that   he
should   look for.   First,   there is   sign-on   and
"converse".   That is,   request use   of the computer
and enter parameters for the run,   such   as sample
number, instrument status,   number of   data points
desired, mode of operation   desired   for this run,
etc.   The parameters, entered at a "console"

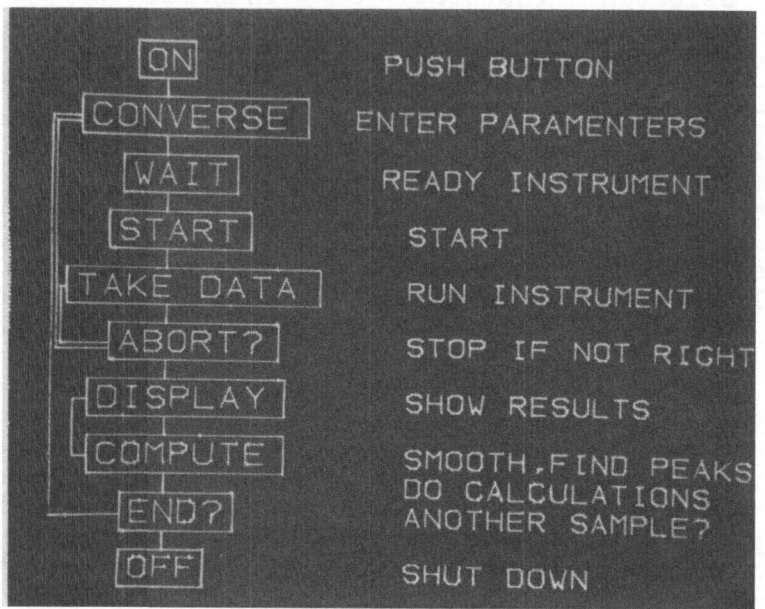

Figure 5.   Software Structure

at   the   instrument   will   be   used   in   report
preparation,   or are   left in   COMMON to tell   his
operating program what to do.   Then   usually it is
a good thing to provide   for a "wait".   Especially
if the sample needs to be injected, or some   other
instrument   preparation performed.   Then a   start
signal   can be given, which tells the computer   to
proceed with the "data taking".   The   data- taking
program will grow   to be as complex   as the user's
skill.   It   will   do   calibration,   take   data,
cross-check,   detect error   conditions,   look   for
inconsistencies, etc. A very important option for
the operator is often overlooked at first, namely,
a way to "stop" the operation at any time.   Don't
pull the   plug   on   the   computer⸘ Do   provide a
section   of   program   which   examines   a   switch
occasionally   to see   if   the   operator   wishes to
continue,   or   abort   the   run.   Data   processing
programs,   and display   programs,   are   then called
into play, as enough data is accumulated, or after
all data is accumulated.

The question of how much data to take, and
keep, is an experiment design problem that is also
often not properly addressed. Aside from the
question of how much data to take to produce
statistically acceptable conclusions, a problem
often faced is whether to analyze data on-the-fly,
or take all data, and analyze after. Generally,
if core or disk space is available, better results
are obtained if data is analyzed after all of it
is taken. The obvious question of "why take data
between peaks?" is best answered by showing that
you know you are "between peaks" if you have all
the data in the computer and use fitting
techniques, rather than using derivative,
predictive techniques on too-small samples of data
as it comes in. This is analogous to looking at
the whole strip chart recording at once, to see
the peaks, as opposed to cutting it up in small
vertical strips, and looking at a single strip to
decide if you are in a peak reigon. There are
many exceptions to such "generalizations", of
course. In many experiments you know roughly
where the peaks of interest are, or the background
is quite flat, so that a discrimination level can
be set, or there is just so much data it can't be
stored, as in high resolution mass spec.

As shown in Figure 4, a graphical output
device, such as a storage scope, is being used
more and more as part of the "operator console" at
the instrument. The reason for this is that it
permits interactive processing of the data, as
well as displays of the data in curve form, rather
than list (printed) form. First of all we, as
human beings, to whom the computer is to present
results, prefer to absorb information in curve
form, rather than numerical list form. This is
the fastest way for the computer to converse with
us. Secondly, we like to see at each step what
the processing routines are doing to the data.
First, we may wish to see the raw data; then, if
we have chosen a smoothing routine, overlay with
the computer smoothed data. Acceptable? OK,
proceed. If we use a background estimation
routine, let the computer draw in the background

as it determined it. Acceptable? OK, proceed.
Subtract the background. Now pick the peaks. Let
the computer show us which peaks it picked; which
it said were multiple peaks, which it
deconvoluted. Acceptable? If not, either rewrite
the peak picking routine on the basis of improved
strategies, or, at the key board, instruct the
computer to remove peak 3 from the list for
further processing; or that it should treat peak 5
as having a small hump on its backside. Thus, the
scientist himself still makes these ultimate
judgements about the data, using his brain and
training in the best way, and not burdened with
the number handling. Figures 6 and 7 show
polaroid pictures of a storage scope display used
in this fashion of interactive calculations.
Beyond this point, the data processing proceeds to
the extraction of information by further
calculations; to comparing with previous spectra;
to entering into an accumulating file information
about this material, to report preparation which
passes on to the next user of our results. The
more of these jobs we begin to include in the
computer, the more we need to draw on the total
computer resources of our institutions, and the
more effectively we can use our own skills to
solve our problems.

How can our local, real time, highly
responsive requirements in our individual
laboratories be joined with the large scale,
multi-tasks requirements of the computing center?
That, I believe, is the main problem facing
proponents of lab automation today, and represents
the proper growth path to keep in mind as we take
steps to use computers on any level.
Symbolically, this connection is represented by
the block figures of Figure 8. There are
certainly areas where the economy of scale in the
center pays off:

  1. Program preparation, in higher level
      languages, for our analysis work and even
      for preparing programs for use on our
      local computers.

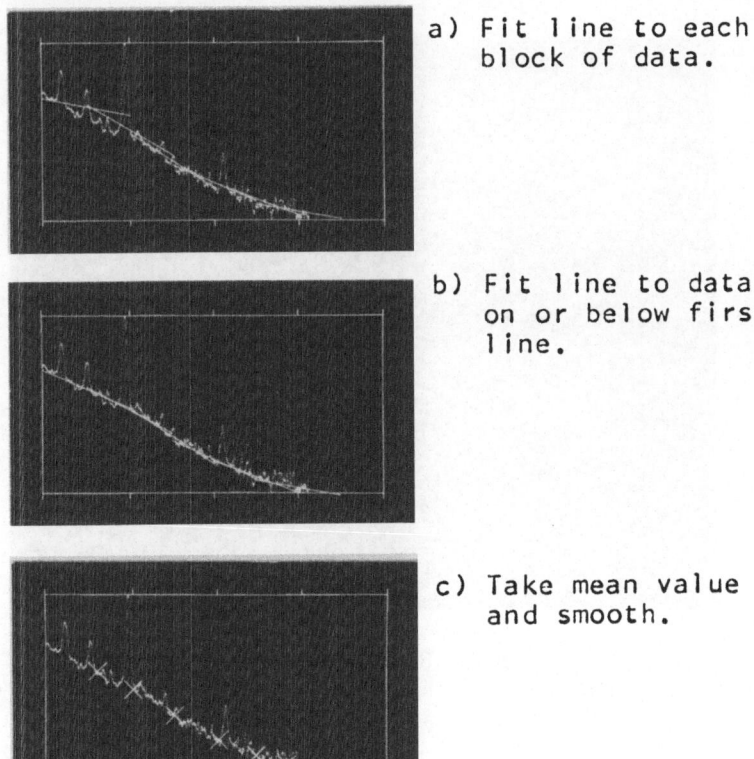

a) Fit line to each
   block of data.

b) Fit line to data
   on or below first
   line.

c) Take mean value
   and smooth.

Figure 6.   Interactive Computing

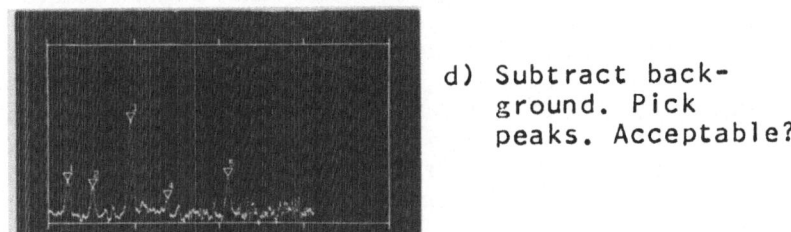

d) Subtract back-
   ground. Pick
   peaks. Acceptable?

Figure 7.   Computer display, analyst's decisions

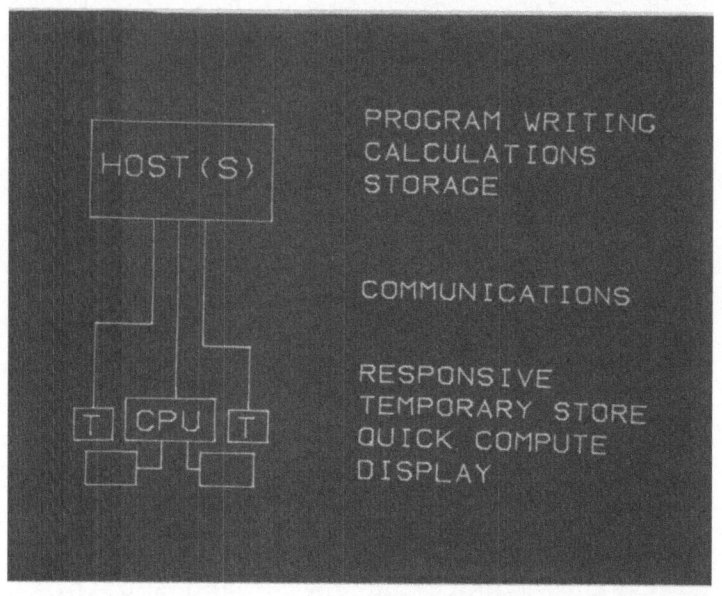

Figure 8.   Distributed System

2.   Storage of our programs and larger data
      files.

3.   And,   obviously,   for   any   larger
      calculations required.

Given the resources to   do larger scale
calculation, we can increase the significance of
our experimental results by improving on the
"approximations" made at our current level of
operation.   (Better "model" of the physics or
chemistry going on in our instrument).   On the
other hand, we can't ask an expensive, large
resource to keep itself free enough to respond to
our immediate instrument operation needs around
the clock.

For that, two or three experimentalists may share a local computer (avoiding some of the political problems of handling 20 users) for their real time needs. The key system analysis problem then becomes how to get these two "unlike" computer resources to respond to each other properly and permit a proper partitioning of the total job that we scientists wish to use the resources for. We are, at the moment, attempting to find some of the answers to this question. In the meantime, whatever the size of the steps taken in the direction of "automation", I feel that it is clear that this growth path should be kept in mind, and the steps made are consistent with an overall pattern of needs.

Adsorption, effect in trace analysis, 12
Air, analysis for pesticides, 133
Air sampling system for pesticides, 143
Aldrin
  concentration in air, 149
  concentration by date, 150
Algae
  cesium concentration factor for, 57
  copper concentration factor for, 56
  filamentous, manganese concentration factor for, 53
Algal growth, effect on trace analysis of water, 14
Aluminum, health effects of, 97
Ammonia nitrogen, methods of analysis, 262
Analytical scheme for treating field samples, 143
Analytical standards, occurrence of contaminants in, 155
Anions
  methods of analysis in water, 201

nutrient, analytical techniques for in water, 241
Antibiotics, analytical criteria for certification, 278
Antimony, health effects of, 95
Arochlors
  composition, 109
  GLC profiles, 116
Arsenic, health effects of, 97
Atomic absorption methods, indirect, for anions, 207, 208
Autoanalyzer
  analytical problems, 337
  use for nitrogen analysis, 265

Bacterial growth, effect on trace analysis of water, 14
Barium, health effects of, 97
Beryllium, health effects of, 95
Bioaccumulation, definition in trace analysis ecosystems, 44
Bioconcentration, definition in trace analysis of ecosystems, 44

Biological factors, effect
  in trace analysis
  of ecosystems, 46
Biomagnification, defini-
  tion of in trace
  analysis of ecosystems,
  45
Biphenyls, polychlorinated,
  interference with
  pesticide analysis, 109
Birds, pesticide concentration factors for, 50
Bivalves
  cesium concentration
    factor for, 57
  pesticide concentration
    factors for, 50
Boats, aluminum, use in C,
  H, and N analysis, 322
Body burden of elements for
  reference man, 72-73
Borate in water, evaluation
  of methods for, 217

Cadmium
  effect on systolic blood
    pressure of rats, 91
  health effects, 90
  summary data, in man, 77
Calculators, comparison with
  computers, 293
Carbon analyses, variation
  with voltage, 330
C, H, and N calculations,
  correction of, 330
Cesium, concentration
  factors in Hudson River
  food chain, 57
Chelating agents in environmental water
  effects in trace analysis,
    13
  stability constants, 92
Chick edema, causative
  agents in, 163

Chick edema factor,
  composition, 163
Chlorinated hydrocarbons,
  biomagnification of, 49
Chloroacne, occurence in
  chemical workers, 158
Chlorophenols
  effects of heating, 165
  occurrence of dibenzo-p-
    dioxins in, 157
Chromium, effects of
  deficiency of, 98
Cobalt oxide, preparation for
  combustion tubes, 321
Coils, mixing, analytical
  problems in autoanalysis,
  341
Collection efficiencies for
  pesticides, 140
Combustion tube packing,
  preparation, 321
Computer, criteria for
  application to analysis,
  287
Computer use in analysis, 361
Computers, comparison with
  calculators, 293
Concentration factor, definition of in trace analysis
  of ecosystems, 44
Copper, concentration factors
  in marine biota, 56
Cortisone acetate tablets,
  variation of content of,
  281
Crabs, cesium concentration
  factor for, 57
Crustacea, cadmium concentrations in, 59
Cyanide in water, evaluation
  of methods for, 218
2,4-D
  concentration in air, 149
  concentration by date, 150
  occurrence of polychloro-
    dibenzo-p-dioxins in, 167

Data acquisition, on-line
    used for autoanalyzers,
    295
    used for automatic plate
    readers, 296
DDD, concentration by date,
    150
DDE, concentration in air,
    149
DDT
    concentration in air, 149
    concentration by date,
    150
    technical, occurrence of
    impurities in, 156
Desipramine, automated
    analysis of, 346
Dicamba, occurrence of
    polychlorodibenzo-p-
    dioxins in, 167
Digestor, automatic use
    in autoanalyzer, 342
Digitoxin preparations,
    variation of content
    of, 282
Digoxin tablets, variation
    of content of, 281
Dioxins, contaminant by-
    products in manufac-
    ture of chlorinated
    phenols, 157
Distillation for separa-
    tion of anions, 204
Drinking water, effects on
    mice of low doses of
    trace elements in,
    86-87
Dual channel flame photo-
    metric detector, 181

Eagle, white-tailed,
    residues of PCB in,
    112
Earth's crust, metals in,
    84-85

Electron capture gas chroma-
    tography, use for
    chlorinated compounds,
    168
Endrin, concentration by date,
    150
Essential trace elements,
    effects of deficiencies,
    98
Ethinyl estradiol tablets,
    variation of content of,
    282
Extraction, liquid-liquid,
    for separation of anions,
    203

Fallout samples, analytical
    procedure, 23
Fallout sampling stations,
    location, 21
Fenthion
    liquid chromatogram, 190
    structure, 188
Filter, continuous, analytical
    problems in autoanalysis,
    341
Fish
    cadmium concentrations in, 59
    cesium concentration factor
    for, 57
    copper concentration factor
    for, 56
    levels of PCB in, 118
    pesticide concentration
    factors for, 50
Flame photometric detection,
    use in residue analysis,
    175
Fluoride
    analysis in natural water,
    229
    colorimetric methods for,
    205
Fluorimetric methods, indirect,
    for anions, 209

Food extracts
  P chromatograms, 179
  S chromatograms, 180
Food web distributions in
  aquatic ecosystems,
  48

Gas chromatography of
  pesticide sample
  extracts, 142
Gas chromatography para-
  meters for pesticide
  analysis, 145
Germanium, health effects
  of, 97
Guillemot, residue of PCB
  in eggs, 112

Halides in water, evalua-
  tion of methods, 219
Heptachlor
  concentration in air, 149
  concentration by date,
  150
Herbicide 2,4,5-T, occur-
  rence of 2,3,7,8-
  tetrachlorodibenzo-p-
  dioxin in, 158
Herring, residues of PCB
  in, 112
Hudson River
  concentration factors
  for cesium in, 57
  concentration factors for
  manganese in, 53
Hydropericardium, cause in
  chicks, 163

IBM 188 computer
  used for autoanalyzers,
  295
  used for automatic plate
  readers, 296
  used for pharmaceutical
  and clinical data,
  298

IBM 1620 computer, use in
  analysis, 358
Imipramine, autoanalysis, 346
Infotronics, digital voltmeter model, 306, 319
Insects, cadmium concentrations
  in, 59
Instrumentation, computerized,
  359
Intake of elements by
  reference man, 72-73
Invertebrates, copper concentration factor for, 56
Iodine, summary data in man,
  78
Ion exchange for separation
  of anions, 204
Ion sensitive electrode,
  analysis for fluoride
  in natural water, 229
Iron, analysis of in sea
  water, 6

Killifish, manganese concentration factor for, 53
Kimwipe, trace element content
  of, 12
Kinetic factors, effects in
  trace analysis of eco-
  systems, 46
Known addition method, used
  in analysis of natural
  waters for fluoride, 233

Lead
  concentration in surface
  air, 26, 27, 28, 29, 30,
  31, 35, 36
  health effects of, 86
  summary data in man, 76
Liquid chromatography, use for
  P and S pesticide
  analysis, 185

Magnesium oxide, preparation
  for combustion tubes, 321

Malathion
  concentration in air, 149
  concentration by date,
    150
  occurrence of a homolog
    in, 156
Man, body composition, 65
Manganese concentration
    factors and food web
    relationship, 54
  concentration factors
    for, 53
Marine biota, copper
    concentration factors
    for, 56
Mass spectrometry
  used for chlorinated
    compounds, 168
  used in identifying
    PCB's, 117
Matrix effect, effect on trace
    analysis of water, 14
Mercury
  health effects, 95
  summary data in man, 75
Metals
  amounts in earth's crust,
    84-85
  amounts in reference man,
    84-85
  amounts in sea water,
    84-85
  approximate annual
    consumption in U.S.,
    84-85
  diseases from excess,
    84-85
  slightly toxic, health
    effects of, 96
Methyl mercury, distribution
    in various organisms,
    51-52
Methyl parathion
  concentration in air, 149
  concentration by date, 151

Mice, effects of low doses of
    trace elements in drink-
    ing water, 86-87
Micro-densitometer, computer
    controlled, 365
Microorganisms, effects on
    volatility of mercury
    from plasma, broth, and
    urine, 13
Microsorban filter, used in
    collecting surface air
    samples, 19
Millipore, trace element
    content of, 12
Mussel, residues of PCB in,
    112

Nickel, health effects of, 95
Nitrate nitrogen, methods of
    analysis, 264
Nitrate in water, evaluation
    of methods, 214
Nitrite in water, evaluation
    of methods, 215
Nitrite nitrogen, methods of
    analysis, 263
Nitrogen
  methods of analysis, 261
  occurrence in the environ-
    ment, 257
Nitrogen cycle, 256

Organochlorine pesticide
    analysis, interference
    by PCB, 113
Orthophosphate, methods of
    analysis for, 253
Output of elements by
    reference man, 72-73
Oxyphenbutazone, autoanalysis
    of, 343
Oysters, effect of copper
    concentration on, 56

Palladium, health effects
     of, 97
Parathion
  concentration in air, 149
  concentration by date,
     150
PCB
  analytical errors caused
     by, 120
  quantitation in analysis,
     124
  separation in analysis,
     121
Pentachlorophenol, occur-
     rence of polychloro-
     dibenzo-p-dioxins in,
     167
Perkin-Elmer model 240, C,
     H, and N analyzer, 319
  modification of, 324, 326
Pesticide residues contain-
     ing phosphorus and
     sulfur, analysis of,
     175
Pesticide standards, gas
     chromatograms of, 178
Pesticides
  airborne concentration by
     date, 151
  airborne, concentration in
     field samples, 149
  airborne, detection limits,
     148
  airborne, monitoring, 135
  airborne, sampling system
     for, 137
  analysis in air, 133
  concentration factors for,
     50
  contaminants in, 153
  environmental distribution
     mechanisms, 133
  response in dual channel
     flame photometric
     detection, 182-183
  response ratios of, 184

Phenmetrazine hydrochloride,
     automated analysis of,
     346
Phenylbutazone, autoanalysis
     of, 343
Phosphates
  analysis in water, 245
  chemical differentiation
     in water, 249
  physical fractionation in
     water, 248
Phosphorus, occurrence in the
     environment, 242
Plankton
  cadmium concentrations in,
     59
  cesium concentration factor,
     57
  manganese concentration
     factor, 53
  pesticide concentration
     factors, 50
Plants
  cadmium concentrations in,
     59
  cesium concentration factor,
     57
  pesticide concentration
     factors, 50
Plexiglas, trace element
     content, 12
Phosphate, in water, evalua-
     tion of methods for, 210
Phosphorus, detection in
     pesticide analysis, 175
Polychlorinated biphenyls
     (PCB)
  properties, 111
  residues, 112
  structure, 110
  uses, 114
Polyethylene, trace elements
     content of, 12
Potentiometry, use in analysis
     of sea water for fluoride,
     231

Pumps, proportioning, analytical problems of, 340

Pyrex, trace element content of, 12

PVC, trace element content of, 12

p-Values, of pesticides, 186-187

Quartz, trace element content of, 12

Rats, effects of low doses of trace elements in drinking water, 89

Reagents, contamination by trace elements, 11

Recovery efficiencies of pesticides, 144

Reference man, 67

Response ratios of pesticides, 184

Rhodium, health effects of, 97

Roots Connersville Blower, used in collecting surface air samples, 19

Rubber, trace element content, 12

Sample ladle
  automation of, 327
  quartz, use of in C, H, and N analysis, 323

"Sampler II," analytical problems in auto-analysis, 339

Sampling, effect on trace analysis of water, 15

Seal, residue of PCB in, 112

Sea Water
  analysis of iron in, 6
  metals in, 84-85
  trace element content, 12

Sediments, pesticide concentrations in, 50

Sephadex G15 and G25, used in separating phosphorus compounds, 255

Silicate in water, evaluation of methods for, 213

Silver oxide, preparation for combustion tubes, 321

Silver tungstate, preparation for combustion tubes, 321

Silver vanadate, preparation for combustion tubes, 321

Silvex, occurrence of polychlorodibenzo-p-dioxins in, 167

Slope method, used in analysis of natural waters for fluoride, 237

"Solid prep," analytical problems in autoanalysis, 338

Spectrophotometer, analytical problems in autoanalysis, 342

Spectrophotometric methods, indirect, for anions, 207

Standard man, 66

Striped bass, manganese concentration factor for, 53

Sulfate in water, evaluation of methods for, 215

Sulfur, detection in pesticide analysis, 175

Surface air samples, analytical procedure, 22

Surface air sampling stations, location, 20

Surfactants, anionic, in water, evaluation of methods for, 220

Teflon, trace element content of, 12

Tellurium, health effects of, 97

Time sharing, commercial, used for gas chromatography, 304

Tin, health effects of, 97

Titanium, health effects of, 97

Trace metal burdens in man, 63

Trace metals
carcinogenicity, 99-100
concentration in precipitation in New York City, 33-38
concentration in precipitation in the U.S.A., 38
concentration in total fallout in New York City, 32

Trace methods, reliability of, 3

2,3,7,8-Tetrachlorodibenzo-p-dioxin (TCDD), causative agent for chloroacne, 158

Tetrachlorophenol, occurrence of polychloro-dibenzo-p-dioxins in, 167

Titration microcoulometric for nitrogen, 265

2,4,5,Trichlorophenoxy acetic acid (2,4,5,T)
occurrence of polychlorodi-benzo-p-dioxins in, 167
occurrence of 2,3,7,8 tetrachlorodibenzo-p-dioxin (TCDD) in, 158
synthesis, 159
table of impurities in, 162

Trichlorophenol, occurrence of polychlorodibenzo-p-dioxins in, 167

Tungsten, health effects of, 97

Variability, sources in trace metal analysis, 7

Voltage standard, internal, for C, H, and N analyzer, 328

Wang calculator model 360, 319

Water, analysis of trace metals in, 5

Worms, cadmium concentrations in, 59

X-ray diffractometer, computerized, 358

Zinc, effects of deficiency of 98